NEW OPTIMIZATION ALGORITHMS AND THEIR APPLICATIONS

NEW OPTIMIZATION ALGORITHMS AND THEIR APPLICATIONS

Atom-Based, Ecosystem-Based and Economics-Based

WEIGUO ZHAO

Professor, School of Water Conservancy and Hydroelectric Power, Hebei University of Engineering, China

LIYING WANG

Professor, School of Water Conservancy and Hydroelectric Power, Hebei university of Engineering, China

ZHENXING ZHANG

Hydrologist, University of Illinois at Urbana-Champaign, USA

ELSEVIER

Science Press
Beijing

Elsevier
Radarweg 29, PO Box 211, 1000 AE Amsterdam, Netherlands
The Boulevard, Langford Lane, Kidlington, Oxford OX5 1GB, United Kingdom
50 Hampshire Street, 5th Floor, Cambridge, MA 02139, United States

Library of Congress Cataloging-in-Publication Data
A catalog record for this book is available from the Library of Congress

British Library Cataloguing-in-Publication Data
A catalogue record for this book is available from the British Library

ISBN: 978-0-323-90941-9

This book was originally published in China with the ISBN 978-7-03-064198-4

Publisher: Glyn Jones
Editorial Project Manager: Naomi Robertson
Production Project Manager: Prasanna Kalyanaraman
Cover Designer: Miles Hitchen

Typeset by SPi Global, India

Contents

Preface

Optimization means to search for optimal solutions effectively and efficiently from a given solution space under the given constraints, by either maximizing or minimizing the objective function. Metaheuristic optimization algorithms, which are powerful tools for addressing these challenging optimization problems, are becoming increasingly popular. For this book, three novel intelligent optimization techniques, based on atomic dynamics, artificial ecosystems, and economics were developed, tested, and applied to practical engineering issues. The book is organized as follows:

Chapter 1 introduces the development of intelligent optimization algorithms and their advantages and disadvantages.

Chapter 2 introduces the process and basic principles of the atom search optimization algorithm and tests the algorithm.

Chapter 3 applies the atom search optimization algorithm in engineering issues.

Chapter 4 introduces the proposed process and basic principle of the artificial ecosystem-based optimization algorithm and tests the algorithm.

Chapter 5 employs the artificial ecosystem-based optimization algorithm in engineering applications.

Chapter 6 presents the inspiration and description of the supply-demand-based optimization.

Chapter 7 demonstrates the applications of the supply-demand-based optimization in different engineering cases.

Appendix A lists mathematical benchmark functions used to test the performance of various algorithms.

Appendix B lists optimization formulations of various engineering problems.

Appendix C provides the MATLAB codes of three algorithms.

We have been working on intelligent optimization algorithms for many years and have made some achievements. The book is compiled on the basis of the results of these achievements.

We appreciate the support of the administration and colleagues of the School of Water Resources and Hydropower, Hebei University of Engineering, China, as well as professors from other universities. We would like to express our respect and gratitude for the references in the book. We also acknowledge the publishers who permitted our papers to be adopted and utilized in the preparation of this book.

This book can be used by technicians who are engaged in computing and intelligent algorithms and various engineering practices that use optimization. It is also a good reference for postgraduates and teachers of relevant specialties in colleges and universities.

Our warmest gratitude is due to our families for their continued support in the course of preparing this book. Dr. Zhao worked on the ideas and papers, which motivated the preparation of this book, in Illinois State Water Survey, University of Illinois at Urbana-Champaign, USA. Preparing the book brought back all the beautiful memories in Champaign, IL. Owing to the limitation of our knowledge, we are not able to eradicate incompleteness or errors in the book. Your suggestions would be much appreciated. Please contact us with your suggestions via the publisher.

Weiguo Zhao
Liying Wang
Zhenxing Zhang

Acknowledgments

Acknowledgment is given to Springer for permission to reprint material from the following paper:

Zhao W, Wang L, Zhang Z. 2020. Artificial ecosystem-based optimization: A novel nature-inspired meta-heuristic algorithm, Neural Computing and Applications 32:9383–9425.

Acknowledgment is given to the Institute of Electrical and Electronic Engineers (IEEE) for permission to reprint a portion of the material from the following paper:

Zhao W, Wang L, Zhang Z. 2019. Supply-demand-based optimization: A novel economics-inspired algorithm for global optimization, IEEE Access 7:73182–73206. © [2019] IEEE.

Acknowledgment is also given to Elsevier for permission to reprint material from the following papers:

Zhao W, Wang L. 2016. An effective bacterial foraging optimizer for global optimization, Information Sciences 329:719–735.

Zhao W, Wang L, Zhang Z. 2019. Atom search optimization and its application to solve a hydrogeologic parameter estimation problem, Knowledge-Based Systems 163:283–304.

The authors are grateful that the copyright permission was granted from the above publishers.

CHAPTER 1

Introduction

Contents

1.1 Optimization algorithms

Optimization algorithms are increasingly popular in intelligent computing and are widely applied to a large number of real-world engineering problems. Their popularity derives from the following aspects. First, all of these optimization techniques have some fundamental theories and mathematical models that have been proved to be reasonable, which come from the real world and are inspired by all types of physical phenomena or biological behaviors (Kirkpatrick et al., 1983; Kennedy and Eberhart, 1995). The theories about these optimization algorithms are simple and easy to understand. Second, these optimization algorithms can be thought of as a black box. This means that given a set of inputs, these algorithms can easily provide a set of outputs for any optimization problem. They are very flexible and versatile because one can change the structures and parameters of the algorithms to obtain better solutions. Third, metaheuristic algorithms can effectively avoid local optima, which is very valuable for addressing engineering problems which are typically considered as multimodal functions. In addition, one can develop their variants by absorbing the merits of other algorithms to improve the accuracy of solutions within a reasonable time. Fourth, the metaheuristic optimization algorithms can tackle different types of problems including, but not limited to, single-objective and multiobjective problems, low-dimensional and high-dimensional problems, unimodal and multimodal problems, and discrete and continuous problems (Liu et al., 2017; Li et al., 2018; Duan et al., 2018; Zhao et al., 2019).

New Optimization Algorithms and their Applications
https://doi.org/10.1016/B978-0-323-90941-9.00001-6

1.2 A short outline of optimization algorithms

Since the 1970s, many optimization algorithms have been developed and applied to different optimization problems. These algorithms, which mimic natural or physical phenomena, have provided effective and robust techniques for solving complex optimization problems in a wide spectrum of disciplines. Many metaheuristic algorithms with different inspirations have been proposed and successfully used in a variety of fields, which are roughly classified into four categories (Hare et al., 2013): evolution-inspired, (Mühlenbein et al., 1988; Gong et al., 2013, 2014), physics-inspired (Geem et al., 2001), swarm-inspired (Krause et al., 2013), and human-inspired (Montiel et al., 2007).

Evolution-inspired algorithms are a stochastic, population-based approach. Protecting a population's diversity is very important for the sustainable development of the algorithms iteratively. Many evolution-inspired algorithms maintain a population's diversity by mimicking basic genetic rules, including reproduction, mutation, selection, chemotaxis, elimination, and migration (Passino, 2002; Falco et al., 2012). These algorithms randomly initialize a population evolved from subsequent iterations and evaluate the individual quality using a fitness function. Genetic algorithm (GA), originally presented by Holland (1975), is a well-known classic evolutionary algorithm (EA). As GA can generally obtain high-quality solutions using mutation, crossover, and selection steps, the original version and its variants are widely applied to many real-world problems (Gong et al., 2018). Since its emergence, a series of schemes aiming to enhance GA have been developed. With increasing popularity of GA, a number of other evolution-based algorithms in the literature, including evolutionary strategies (ES) (Beyer and Schwefel, 2002), differential evolution (DE) (Rocca et al., 2011), evolutionary programming (EP) (Juste et al., 1999), and memetic algorithms (MA) (Moscato et al., 2007), have been proposed. Additionally, many types of new EAs have been proposed recently, such as bacterial foraging optimization (BFO) (Passino, 2002), bat algorithm (BA) (Yang and Hossein, 2012), fruit fly optimization algorithm (FOA) (Pan, 2012), monkey king evolutionary (MKE) (Meng and Pan, 2016), artificial algae algorithm (AAA) (Uymaz et al., 2015), biogeography-based optimization (BBO) (Simon, 2009), yin-yang-pair optimization (YYPO) (Punnathanam and Kotecha, 2016), invasive weed optimization (IWO) (Mehrabian and Lucas, 2006), and dynamic virtual bats algorithm (DVBA) (Topal and Altun, 2016).

Physics-inspired algorithms simulate physical laws in the universe, among which, simulated annealing (SA) (Kirkpatrick et al., 1983) is one of the most well-known algorithms. SA is inspired from the annealing process used in physical material in which a heated metal cools and freezes into a crystal texture with a minimum amount of energy. Recently, many novel physics-inspired algorithms have been proposed, such as gravitational search algorithm (GSA) (Rashedi et al., 2009), electromagnetism-like mechanism (EM) algorithm (Birbil and Fang, 2003), particle collision algorithm (PCA) (Sacco and De Oliveira, 2015), vortex search algorithm (VSA) (Doğan and Ölmez, 2015), water evaporation optimization (WEO) (Kaveh and Bakhshpoori, 2016), space gravitational algorithm (SGA) (Hsiao et al., 2005), big bang-big crunch algorithm (BB-BC) (Genç et al., 2010), galaxy based algorithm (GBA) (Shah-Hosseini, 2009), big crunch algorithm (BCA) (Kripka and Kripka, 2008), integrated radiation algorithm (IRA) (Chuang and Jiang, 2007), water drops algorithm (WDA) (Shah-Hosseini, 2009), charged system search (CSS) (Kaveh and Talatahari, 2010), magnetic optimization algorithm (MOA) (Mirjalili and Hashim, 2012), gravitation field algorithm (GFA) (Zheng et al., 2010), ions motion algorithm (IMA) (Javidy et al., 2015), water wave optimization (WWO) (Zheng, 2015), gravitational interactions optimization (GIO) (Flores et al., 2011), teaching-learning-based optimization (TLBO) (Rao et al., 2012), hysteretic optimization (HO) (Zarand et al., 2002), thermal exchange optimization (TEO) (Kaveh and Dadras, 2017), light ray optimization (LRO) (Shen and Li, 2009), heat transfer search (HTS) (Patel and Savsani, 2015), spiral optimization algorithm (SOA) (Tamura and Yasuda, 2011), water cycle algorithm (WCA) (Eskandar et al., 2012), and curved space optimization (CSO) (Moghaddam et al., 2012).

Swarm-inspired algorithms mimic the collective behaviors of self-organization and shape-formation, be it natural or artificial (Beni and Wang, 1993). There are two classic swarm-inspired algorithms. One is particle swarm optimization (PSO) (Kennedy and Eberhart, 1995), which mimics bird-flocking behaviors. In PSO, every agent moves around the search space to improve its solution, and their personal best positions and the globally best position found so far are reserved, by which their positions are updated locally and socially. The other is ant colony optimization (ACO) (Dorigo et al., 1996), which follows the foraging process of an ant colony. Essentially, ants communicate with each other by pheromone trails through path formations, which assist them in finding the shortest path between the nest and food source. There are many newly developed swarm-inspired algorithms, such as artificial bee colony (ABC) (Akay and Karaboga, 2012), salp swarm algorithm (SSA) (Mirjalili et al., 2017), krill herd algorithm (KH) (Gandomi and Alavi, 2012), tree-seed algorithm (TSA) (Kiran, 2015), social spider optimization (SSO) (Cuevas et al., 2013), bird mating optimizer (BMO) (Askarzadeh, 2014), cuckoo search (CS) (Yang and Deb, 2019), grasshopper optimization algorithm (GOA) (Saremi et al., 2017), sine-cosine algorithm (SCA) (Mirjalili, 2016), moth swarm algorithm (MSA) (Mohamed et al., 2017), dolphin echolocation (DE) algorithm (Kaveh and Farhoudi, 2013), hunting search (HS) algorithm (Oftadeh et al., 2010), migrating birds optimization (MBO) (Duman et al., 2012), firefly algorithm (FA) (Yang, 2010), monkey search (MS) algorithm (Mucherino and Seref, 2007), and squirrel search algorithm (SSA) (Jain et al., 2018).

Human-inspired algorithms are a recently developed category in intelligence computing. They mathematically stimulate social activities and ideology in humans to find near-optimal solutions. The society and civilization algorithm (SCA) (Ray and Liew, 2003) is a typical representative of human-inspired algorithms. SCA imitates the intra and social interactions within a formal society and the civilization model. A society corresponds to a set of mutually interacting individuals and a civilization is a set of all such societies. All individuals in each society interact with each other and make improvements under the guidance of a leader belonging to the same society. Meanwhile, each leader interacts with leaders of other societies to migrate to a developed society. This leader migration mechanism helps SCA globally search promising regions in the variable space. Additionally, SCA is advantageous in dealing with constrained optimization problems owing to the leader identification mechanism. Some of other human-inspired algorithms include league championship algorithm (LCA) (Kashan, 2009), social group optimization (SGO) (Satapathy and Naik, 2016), social emotional optimization algorithm (SOA) (Xu et al., 2010), socio evolution and learning optimization algorithm (SELO) (Kumar et al., 2018), ideology algorithm (IA) (Huan et al., 2017), and cultural evolution algorithm (CEA) (Kuo and Lin, 2013).

Compared with evolution-inspired or physics-inspired algorithms, swarm-inspired algorithms have some distinctive characteristics. On the one hand, part or all of the historical information about the population needs to be preserved, because every agent depends on the information to determine a new position in the search space over subsequent iterations. However, evolution-inspired algorithms require more operators. Swarm-inspired algorithms generally update positions of the population by interaction rules as standard formulas. On the other hand, swarm-inspired algorithms generally have two behaviors: exploration and exploitation (Alba and Dorronsoro, 2005; Lynn and Suganthan, 2015). Exploration means the ability of the algorithms to search for new solutions far from the current solution in the entire search space. Exploitation means the ability of the algorithms to search for the best solution near a new solution that has already been found. In such algorithms, the range of every agent in the search space is scaled to a consensus in its neighborhood, and agents randomly explore the whole search space. If an agent or its neighbors find a good region, this region will be intensively exploited. Otherwise, they still extensively explore other regions, thus indicating their better self-adaptation in searching the global optima. From these perspectives, swarm-inspired algorithms have many advantages over other algorithms. Many evolution-inspired or physics-inspired algorithms have swarm-inspired characteristics, such as PSO, ACO, CS, BFO, GSA, and so on. These algorithms not only reflect the nature of biological phenomena or physical laws, but also share a common characteristic of exploration and exploitation. Thus, they are more competitive than those without swarm-inspired characteristics. However, providing a proper balance between exploration and exploitation will lead to an optimal performance of the algorithms, so it is one of the most important tasks in the development of any stochastic optimization algorithm.

With the development of economy, society, and technology, a great number of complex and challenging optimization problems have accordingly arisen in different fields. As an illustration, the emergence of ride-sharing companies that offer transportation on demand on a large scale, together with the increasing availability of corresponding demand data sets, develops a new complex optimization problem of effective handling of the routing network (Bertsimas et al., 2018). At present, tracking the growing interest in a closed-loop supply chain by both practitioners and academia is easily possible. Many factors, including environmental legislation, customer awareness, and the economical motivations of the organizations, can transform closed-loop supply chain issues into a unique and vital subject in supply chain management. Designing and planning a closed-loop supply chain is performed by an effective optimizer in a reasonable time (Soleimani and Kannan, 2015). Solving the large-scale, highly nonlinear, nonconvex, non-smooth, nondifferential, noncontinuous, and complex combined heat and power economic dispatch (CHPED) problems urgently requires optimization algorithms (Beigvand et al., 2017). Another challenging optimization problem is the identification of pollutant sources for river pollution incidents, which are caused by accident or illegal emissions (Zhang and Xin, 2017).

Although a number of optimization algorithms have been introduced so far, new optimization algorithms are still being developed to tackle emerging complex optimization problems to obtain a better outcome. Furthermore, according to the No Free Lunch Theorem of Optimization (Wolpert and Macready, 1997), there is no one optimization algorithm performing the best for all types of problems. This theorem keeps this research field active and encourages relevant scholars to develop new algorithms for better optimization. Based on the above, three metaheuristic algorithms with swarm characteristics including atom search optimization, artificial ecosystem-based optimization, and supply-demand-based optimization, have been developed and examined using various types of benchmark functions, and then engineering applications have demonstrated their feasibility and effectiveness in tackling real-world problems (Zhao et al., 2019).

1.3 Organization of this book

This book is organized in the following manner:

This chapter introduces the definition of optimization, the development process and the detailed processes to solve the problem.

Chapter 2 is organized as follows. Section 2.2 presents the inspiration of an atom search optimization algorithm; Section 2.3 describes the novel atom search optimization algorithm; Section 2.4 gives a comparative study and discussion on the benchmark functions; and finally, Section 2.5 presents some conclusions and suggests some future research directions.

Chapter 3 is organized as follows. Sections 3.2 and 3.3 describe the application of atom search optimization algorithm for parameter estimation for a chaotic system and the circular antenna array design problems, which demonstrates its availability and effectiveness in solving real-world problems. Section 3.4 presents the spread spectrum radar polyphase code design problem which is compared with its counterparts. The experimental results suggest that atom search optimization is effective and that it is a promising alternative to PSO, GA, and BFO extensively used in many real-world problems.

Chapter 4 introduces a novel nature-inspired metaheuristic optimization algorithm, named artificial ecosystem-based optimization. Section 4.2

describes the algorithm and the concepts behind it in detail. Some mathematical optimization problems are utilized to test the validity of the proposed optimizer form different perspectives in Section 4.3. Section 4.4 gives a conclusion.

Chapter 5 investigates the applications of artificial ecosystem-based optimization to some engineering design cases and the static economic load dispatch problem and hydrothermal scheduling problem.

Chapter 6 presents the inspiration and supply-demand-based optimization in detail, and then the experimental results on a set of mathematical benchmark functions are analyzed.

Chapter 7 demonstrates the effectiveness of supply-demand-based optimization in solving four engineering problems. Finally we conclude the work and suggest several directions for future research.

References

Akay, B., Karaboga, D., 2012. A modified artificial bee colony algorithm for real-parameter optimization. Inf. Sci. 192, 120–142.

Alba, E., Dorronsoro, B., 2005. The exploration/exploitation tradeoff in dynamic cellular genetic algorithms. IEEE Trans. Evol. Comput. 9 (2), 126–142.

Askarzadeh, A., 2014. Bird mating optimizer: an optimization algorithm inspired by bird mating strategies. Commun. Nonlinear Sci. Numer. Simul. 19 (4), 1213–1228.

Beigvand, S.D., Abdi, H., La Scala, M., 2017. Hybrid gravitational search algorithm-particle swarm optimization with time varying acceleration coefficients for large scale CHPED problem. Energy 126, 841–853.

Beni, G., Wang, J., 1993. Swarm intelligence in cellular robotic systems. In: Robots and Biological Systems: Towards a New Bionics? Springer, Berlin, Heidelberg, pp. 703–712.

Bertsimas, D., Jaillet, P., Martin, S., et al., 2018. Online vehicle routing: the edge of optimization in large-scale applications. Oper. Res., 1–20.

Beyer, H.G., Schwefel, H.P., 2002. Evolution strategies-a comprehensive introduction. Nat. Comput. 1 (1), 3–52.

Birbil, Ş.İ., Fang, S.C., 2003. An electromagnetism-like mechanism for global optimization. J. Glob. Optim. 25 (3), 263–282.

Chuang, C.L., Jiang, J.A., 2007. Integrated radiation optimization: inspired by the gravitational radiation in the curvature of space-time. In: Evolutionary Computation, CEC 2007, IEEE Congress on IEEE, pp. 3157–3164.

Cuevas, E., Cienfuegos, M., Zaldívar, D., Pérez-Cisneros, M., et al., 2013. A swarm optimization algorithm inspired in the behavior of the social-spider. Expert Syst. Appl. 40 (16), 6374–6384.

Doğan, B., Ölmez, T., 2015. A new metaheuristic for numerical function optimization: vortex search algorithm. Inf. Sci. 293, 125–145.

Dorigo, M., Maniezzo, V., Colorni, A., et al., 1996. Ant system: optimization by a colony of cooperating agents. IEEE Trans. Syst. Man Cybern. B 26 (1), 29–41.

Duan, P.Y., Li, J.Q., Wang, Y., et al., 2018. Solving chiller loading optimization problems using an improved teaching–learning-based optimization algorithm. Optim. Control Appl. Methods 39 (1), 65–77.

Duman, E., Uysal, M., Alkaya, A.F., et al., 2012. Migrating birds optimization: a new metaheuristic approach its performance on quadratic assignment problem. Inf. Sci. 217, 65–77.

Eskandar, H., Sadollah, A., Bahreininejad, A., et al., 2012. Water cycle algorithm-a novel metaheuristic optimization method for solving constrained engineering optimization problems. Comput. Struct. 110, 151–166.

Falco, I., Della Cioppa, A., Maisto, D., et al., 2012. Biological invasion-inspired migration in distributed evolutionary algorithms. Inf. Sci. 207, 50–65.

Flores, J.J., López, R., Barrera, J., et al., 2011. Gravitational interactions optimization. In: International Conference on Learning and Intelligent Optimization. Springer, Berlin, Heidelberg, pp. 226–237.

Gandomi, A.H., Alavi, A.H., 2012. Krill herd: a new bio-inspired optimization algorithm. Commun. Nonlinear Sci. Numer. Simul. 17 (12), 4831–4845.

Geem, Z.W., Kim, J., Loganathan, G.V., et al., 2001. A new heuristic optimization algorithm: harmony search. Trans. Simul. 76 (2), 60–68.

Genç, H.M., Eksin, I., Erol, O.K., et al., 2010. Big bang-big crunch optimization algorithm hybridized with local directional moves and application to target motion analysis problem. In: Systems Man and Cybernetics (SMC), 2010 IEEE International Conference on. IEEE, pp. 881–887.

Gong, D., Sun, J., Ji, X., et al., 2013. Evolutionary algorithms with preference polyhedron for interval multi-objective optimization problems. Inf. Sci. 233, 141–161.

Gong, D., Ji, X., Sun, J., Sun, X., et al., 2014. Interactive evolutionary algorithms with decision-maker's preferences for solving interval multi-objective optimization problems. Neurocomputing 137, 241–251.

Gong, D., Sun, J., Miao, Z., et al., 2018. A set-based genetic algorithm for interval many-objective optimization problems. IEEE Trans. Evol. Comput. 22 (1), 47–60.

Hare, W., Nutini, J., Tesfamariam, S., et al., 2013. A survey of non-gradient optimization methods in structural engineering. Adv. Eng. Softw. 59, 19–28.

Holland, J.H., 1975. Adaptation in Natural and Artificial Systems. University of Michigan Press, Ann Arbor, MI.

Hsiao, Y.T., Chuang, C.L., Jiang, J.A., et al., 2005. A novel optimization algorithm: space gravitational optimization, systems, man and cybernetics. In: 2005 IEEE International Conference on. IEEE. vol. 3, pp. 2323–2328.

Huan, T.T., Kulkarni, A.J., Kanesan, J., et al., 2017. Ideology algorithm: a socio-inspired optimization methodology. Neural Comput. & Applic. 28 (1), 845–876.

Jain, M., Singh, V., Rani, A., et al., 2018. A novel nature-inspired algorithm for optimization: squirrel search algorithm. Swarm Evol. Comput. https://doi.org/10.1016/j.swevo.2018. 02. 013.

Javidy, B., Hatamlou, A., Mirjalili, S., et al., 2015. Ions motion algorithm for solving optimization problems. Appl. Soft Comput. 32, 72–79.

Juste, K.A., Kita, H., Tanaka, E., et al., 1999. An evolutionary programming solution to the unit commitment problem. IEEE Trans. Power Syst. 14 (4), 1452–1459.

Kashan, A.H., 2009. League championship algorithm: a new algorithm for numerical function optimization. In: 2009 International Conference of Soft Computing and Pattern Recognition. IEEE 2009, pp. 43–48.

Kaveh, A., Bakhshpoori, T., 2016. Water evaporation optimization: a novel physically inspired optimization algorithm. Comput. Struct. 167, 69–85.

Kaveh, A., Dadras, A., 2017. A novel meta-heuristic optimization algorithm: thermal exchange optimization. Adv. Eng. Softw. 110, 69–84.

Kaveh, A., Farhoudi, N., 2013. A new optimization method: dolphin echolocation. Adv. Eng. Softw. 59, 53–70.

Kaveh, A., Talatahari, S., 2010. A novel heuristic optimization method: charged system search. Acta Mech. 213 (3), 267–289.
Kennedy, J., Eberhart, R., 1995. Particle swarm optimization. In: Proceedings of the 1995 IEEE International Conference on Neural Networks, pp. 1942–1948.
Kiran, M.S., 2015. TSA: tree-seed algorithm for continuous optimization. Expert Syst. Appl. 42 (19), 6686–6698.
Kirkpatrick, S., Gelatt, C.D., Vecchi, M.P., et al., 1983. Optimization by simulated annealing. Science 220 (4598), 671–680.
Krause, J., Cordeiro, J., Parpinelli, R.S., et al., 2013. A survey of swarm algorithms applied to discrete optimization problems. In: Swarm Intelligence and Bio-inspired Computation: Theory and Applications. Elsevier Sci. Technol. Books, pp. 169–191.
Kripka, M., Kripka, R.M.L., 2008. Big crunch optimization method. In: International Conference on Engineering Optimization, Brazil, pp. 1–5.
Kumar, M., Kulkarni, A.J., Satapathy, S.C., 2018. Socio evolution & learning optimization algorithm: a socio-inspired optimization methodology. Futur. Gener. Comput. Syst. 81, 252–272.
Kuo, H.C., Lin, C.H., 2013. Cultural evolution algorithm for global optimizations and its applications. J. Appl. Res. Technol. 11 (4), 510–522.
Li, J.Q., Sang, H.Y., Han, Y.Y., et al., 2018. Efficient multi-objective optimization algorithm for hybrid flow shop scheduling problems with setup energy consumptions. J. Clean. Prod. 181, 584–598.
Liu, Y., Gong, D., Sun, J., Jin, Y., et al., 2017. A many-objective evolutionary algorithm using a one-by-one selection strategy. IEEE Trans. Cybern. 47 (9), 2689–2702.
Lynn, N., Suganthan, P.N., 2015. Heterogeneous comprehensive learning particle swarm optimization with enhanced exploration and exploitation. Swarm Evol. Comput. 24, 11–24.
Mehrabian, A.R., Lucas, C., 2006. A novel numerical optimization algorithm inspired from weed colonization. Ecol. Inf. 1 (4), 355–366.
Meng, Z., Pan, J.S., 2016. Monkey king evolution: a new memetic evolutionary algorithm and its application in vehicle fuel consumption optimization. Knowl.-Based Syst. 97, 144–157.
Mirjalili, S., 2016. SCA: a sine cosine algorithm for solving optimization problems. Knowl.-Based Syst. 96, 120–133.
Mirjalili, S.A., Hashim, S.Z.M., 2012. BMOA: binary magnetic optimization algorithm. Int. J. Mach. Learn. Comput. 2 (3), 204.
Mirjalili, S., Gandomi, A.H., Mirjalili, S.Z., Saremi, S., Faris, H., Mirjalili, S.M., 2017. Salp swarm algorithm: a bio-inspired optimizer for engineering design problems. Adv. Eng. Softw. 114, 163–191.
Moghaddam, F.F., Moghaddam, R.F., Cheriet, M., et al., 2012. Curved space optimization: a random search based on general relativity theory. arXiv, 1208–2214. preprint arXiv.
Mohamed, A.A.A., Mohamed, Y.S., El-Gaafary, A.A.M., Hemeida, A.M., et al., 2017. Optimal power flow using moth swarm algorithm. Electr. Power Syst. Res. 142, 190–206.
Montiel, O., Castillo, O., Melin, P., Rodríguez-Díaz, A., Sepúlveda, R., 2007. Human evolutionary model: a new approach to optimization. Inf. Sci. 177, 2075–2098.
Moscato, P., Mendes, A., Berretta, R., et al., 2007. Benchmarking a memetic algorithm for ordering microarray data. Biosystems 88 (1), 56–75.
Mucherino, A., Seref, O., 2007. Monkey search: a novel metaheuristic search for global optimization. AIP Conf. Proc. 953 (1), 162–173.
Mühlenbein, H., Gorges-Schleuter, M., Krämer, O., et al., 1988. Evolution algorithms in combinatorial optimization. Parallel Comput. 7 (1), 65–85.

Oftadeh, R., Mahjoob, M.J., Shariatpanahi, M., et al., 2010. A novel meta-heuristic optimization algorithm inspired by group hunting of animals: hunting search. Comput. Math. Appl. 60 (7), 2087–2098.
Pan, W.T., 2012. A new fruit fly optimization algorithm: taking the financial distress model as an example. Knowl.-Based Syst. 26, 69–74.
Passino, K.M., 2002. Biomimicry of bacterial foraging for distributed optimization and control. IEEE Control. Syst. 22 (3), 52–67.
Patel, V.K., Savsani, V.J., 2015. Heat transfer search (HTS): a novel optimization algorithm. Inf. Sci. 324, 217–246.
Punnathanam, V., Kotecha, P., 2016. Yin-Yang-pair optimization: a novel lightweight optimization algorithm. Eng. Appl. Artif. Intell. 54, 62–79.
Rao, R.V., Savsani, V.J., Vakharia, D.P., et al., 2012. Teaching–learning-based optimization: an optimization method for continuous non-linear large scale problems. Inf. Sci. 183 (1), 1–15.
Rashedi, E., Nezamabadi-Pour, H., Saryazdi, S., et al., 2009. GSA: a gravitational search algorithm. Inf. Sci. 179 (13), 2232–2248.
Ray, T., Liew, K.M., 2003. Society and civilization: an optimization algorithm based on the simulation of social behavior. IEEE Trans. Evol. Comput. 7 (4), 386–396.
Rocca, P., Oliveri, G., Massa, A., et al., 2011. Differential evolution as applied to electromagnetics. IEEE Antennas Propag. Mag. 53 (1), 38–49.
Sacco, W.F., De Oliveira, C.R.E., 2015. A new stochastic optimization algorithm based on a particle collision metaheuristic. In: Proceedings of 6th WCSMO, Brazil, pp. 1–6.
Saremi, S., Mirjalili, S., Lewis, A., et al., 2017. Grasshopper optimization algorithm: theory and application. Adv. Eng. Softw. 105, 30–47.
Satapathy, S., Naik, A., 2016. Social group optimization (SGO): a new population evolutionary optimization technique. Complex Intell. Syst. 2 (3), 173–203.
Shah-Hosseini, H., 2009. The intelligent water drops algorithm: a nature-inspired swarm-based optimization algorithm. Int. J. Bio-Inspir. Comput. 1 (1–2), 71–79.
Shen, J., Li, Y., 2009. Light ray optimization and its parameter analysis. In: Computational Sciences and Optimization, CSO 2009, International Joint Conference on IEEE, pp. 918–922.
Simon, D., 2009. Biogeography-based optimization. IEEE Trans. Evol. Comput. 12 (6), 702–713.
Soleimani, H., Kannan, G., 2015. A hybrid particle swarm optimization and genetic algorithm for closed-loop supply chain network design in large-scale networks. Appl. Math. Model. 39 (14), 3990–4012.
Tamura, K., Yasuda, K., 2011. Primary study of spiral dynamics inspired optimization. IEEE Trans. Electr. Electron. Eng. 6 (S1), S98–S100.
Topal, A.O., Altun, O., 2016. A novel meta-heuristic algorithm: dynamic virtual bats algorithm. Inf. Sci. 354, 222–235.
Uymaz, S.A., Tezel, G., Yel, E., et al., 2015. Artificial algae algorithm (AAA) for nonlinear global optimization. Appl. Soft Comput. 31, 153–171.
Wolpert, D.H., Macready, W.G., 1997. No free lunch theorems for optimization. IEEE Trans. Evol. Comput. 1 (1), 67–82.
Xu, Y., Cui, Z., Zeng, J., 2010. Social emotional optimization algorithm for nonlinear constrained optimization problems. In: International Conference on Swarm, Evolutionary, and Memetic Computing. Springer, Berlin, Heidelberg, pp. 583–590.
Yang, X.S., 2010. Firefly algorithm, stochastic test functions and design optimization. Int. J. Bio-Inspir. Comput. 2 (2), 78–84.
Yang, X.S., Deb, S., 2019. Cuckoo search via Lévy flights, nature & biologically inspired computing. In: NaBIC 2009, World Congress on IEEE, pp. 210–214.

Yang, X.S., Hossein, G.A., 2012. Bat algorithm: a novel approach for global engineering optimization. Eng. Comput. 29 (5), 464–483.
Zarand, G., Pazmandi, F., Pál, K.F., et al., 2002. Using hysteresis for optimization. Phys. Rev. Lett. 89 (15), 150–201.
Zhang, S.P., Xin, X.K., 2017. Pollutant source identification model for water pollution incidents in small straight rivers based on genetic algorithm. Appl. Water Sci. 7 (4), 1955–1963.
Zhao, W., Wang, L., Zhang, Z., 2019. Atom search optimization and its application to solve a hydrogeologic parameter estimation problem. Knowl.-Based Syst. 163, 283–304.
Zheng, Y.J., 2015. Water wave optimization: a new nature-inspired metaheuristic. Comput. Oper. Res. 55, 1–11.
Zheng, M., Liu, G., Zhou, C., et al., 2010. Gravitation field algorithm and its application in gene cluster. Algorithms Mol. Biol. 5 (1), 32.

CHAPTER 2

Atom search optimization algorithm

Contents

2.1 Introduction

Nature contains boundless secrets that are rich and fantastic. It brings a great deal of inspiration to people who can greatly contribute to social development. Intelligent algorithms (IA), dealing with difficult problems in science and engineering, is an important branch inspired from nature. Since the 1970s, a variety of nature-inspired optimization algorithms have been put forward and applied to many real-world problems (Zhang et al., 2008; Poli et al., 2010; Ayman, 2011; Kourakos and Mantoglou, 2013; Zhao and Wang, 2016), all of which consist of two basic characteristics.

First, the algorithms mimic evolving properties and the living habit of biological systems. There are three common algorithms. The genetic algorithm (GA) (Holland, 1975) is a well-known classic optimization algorithm which can generally obtain high-quality solutions using mutation, crossover,

New Optimization Algorithms and their Applications
https://doi.org/10.1016/B978-0-323-90941-9.00002-8

and selection steps, and it is a good global optimization approach. Particle swarm optimization (PSO) (Kennedy and Eberhart, 1995) mimics the social behaviors of a flock of birds. In PSO, every agent moves around the search space to improve its solution, and their personal best positions and the globally best position found so far are reserved by which their positions are updated locally and socially. Ant colony optimization (ACO) (Dorigo et al., 1996), another well-known optimization method, simulates the foraging behaviors of real ant colonies. Essentially, ants communicate with each other by pheromone trails through path formation, which assists them to find the shortest path, signifying a near-optimum solution. With their increasing popularity (Lai et al., 2016; Shahabinejad and Sohrabpour, 2017; Thilak and Amuthan, 2018), quite a number of other similar algorithms have been developed, including evolutionary strategies (ES) (Beyer and Schwefel, 2002), differential evolution (DE) (Rocca et al., 2011), evolutionary programming (EP) (Juste et al., 1999), memetic algorithm (MA) (Moscato et al., 2007), bacterial foraging optimization (BFO) (Passino, 2002), biogeography-based optimization (BBO) (Simon, 2009), cuckoo search (CS) algorithm (Yang and Deb, 2009), artificial bee colony (ABC) (Karaboga and Akay, 2007), and fruit fly optimization algorithm (FOA) (Pan, 2012). It is worth mentioning here that other heuristic approaches inspired from socio-behaviors also have this characteristic, they include the society and civilization algorithm (SCA) (Ray and Liew, 2003), league championship algorithm (LCA) (Kashan, 2009), social-emotional optimization algorithm (SOA) (Xu et al., 2010), teaching-learning-based optimization (TLBO) (Rao et al., 2012), cultural evolution algorithm (CEA) (Kuo and Lin, 2013), soccer league competition algorithm (SLCA) (Moosavian and Roodsari, 2014), social group optimization (SGO) (Satapathy and Naik, 2016), ideology algorithm (IA) (Huan et al., 2017), and socio-evolution and learning optimization (SELO) algorithm (Kumar et al., 2018).

The second basic characteristic of nature-inspired algorithms is that some of them are enlightened from physical laws of different substances, among which, simulated annealing (SA) (Kirkpatrick et al., 1983) is one of the most well-known algorithms. This was inspired by the annealing process used in physical materials in which a heated metal cools and freezes into a crystal texture with a minimum of energy. Along with DE, there are many other similar algorithms developed and successfully performed including electromagnetism-like mechanism (EM) (Birbil and Fang, 2003) based on the attraction-repulsion mechanism of electromagnetism, central force optimization (CFO) (Formato, 2007) and gravitational search algorithm (GSA)

(Rashedi et al., 2009) both inspired by Newton's gravitational law, hysteretic optimization (HO) (Pál, 2006) inspired from demagnetization process, big bang-big crunch (BB-BC) (Erol and Eksin, 2006) inspired from the hypothesis of creation and destruction processes of the universe, and wind driven optimization (WDO) (Bayraktar et al., 2013) based on the earth's atmospheric motion. Despite the emergence of many new optimization approaches, there is no single approach which can perform the best for all types of problems (Wolpert and Macready, 1997).

2.2 Basic molecular dynamics

Atom search optimization (ASO) (Zhao et al., 2019) is inspired by basic molecular dynamics. From the micro perspective, a definition of "matter" based on its physical and chemical structure is: matter is made up of molecules (Barker, 1870). A molecule is the smallest unit of a chemical compound, and it exhibits the same chemical properties as those of that specific compound. A molecule is composed of atoms held together by covalent bonds that vary greatly in terms of complexity and size. Hence all substances are made of atoms and all atoms have mass and volume (Kenkel et al., 2000; Walker and King, 2005). Fig. 2.1 shows the composition of water molecules, each of which is made up of two hydrogen atoms and one oxygen atom, jointly held by two covalent bonds. In an atomic system, all the atoms interact with each other and are in constant motion, whether in the state(s) of gas, liquid, or solid. They are very complex in terms of their structure and microscopic interactions. Because an atomic system is typically composed of

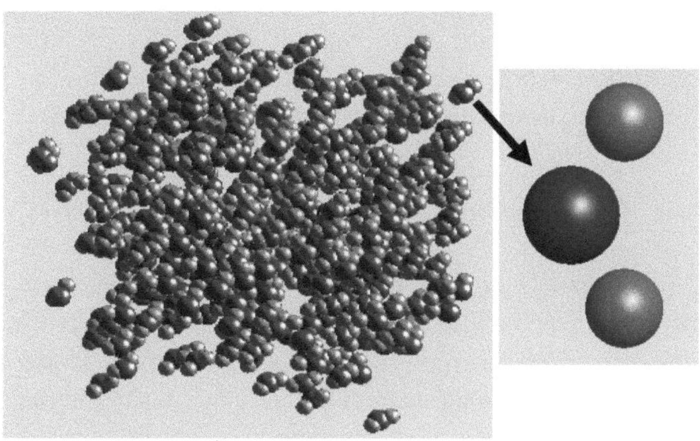

Fig. 2.1 Water molecules and their composition.

numerous atoms, it is analytically impossible to determine the properties that are affected by factors such as temperature, pressure, and so on. With the development of computer technology, molecular dynamics (MD) has rapidly developed in recent years. It circumvents analysis problems by the use of a computer simulation method to examine the physical movements of atoms and molecules.

MD was initially conceived in the field of theoretical physics (Alder and Wainwright, 1959; Rahman, 1964), but its use has been extended to computational chemistry, materials science, and biology. Atomic motion follows classic mechanics (Goldstein et al., 2001). The interaction force among the atoms has two principal characteristics in an atom system. The first is the repulsion to compression, which repels at a close range of crowdedness. The second is the attraction that binds atoms together such as in solid and liquid states. Atoms attract each other over a further range of separation (Rapaport et al., 1996). The potential energy of atoms can account for these two characteristics, and there are a wide variety of pair-wise formulas in the literature used to express the potential energy (Maitland et al., 1981; Gray and Gubbins, 1984). The Lennard-Jones (L-J) potential (Lennard-Jones, 1924; Stone, 1996), initially proposed for liquid, is a simple mathematical model that approximates the interaction force between a pair of atoms. The L-J potential between the ith and the jth atoms is commonly expressed as

$$U(r_{ij}) = 4\varepsilon\left[\left(\frac{\sigma}{r_{ij}}\right)^{12} - \left(\frac{\sigma}{r_{ij}}\right)^{6}\right] \tag{2.1}$$

where ε is the depth of the potential well that represents the strength of the interaction, σ is the length scale that denotes the collision diameter, $\boldsymbol{r}_{ij} = \boldsymbol{x}_j - \boldsymbol{x}_i$, and $\boldsymbol{x}_i = (x_{i1}, x_{i2}, x_{i3})$ is the position of the ith atom in a 3-D space, so the Euclidian distance between the ith and jth atoms is

$$r_{ij} = \|\boldsymbol{x}_j - \boldsymbol{x}_i\| = \sqrt{(x_{i1} - x_{j1})^2 + (x_{i2} - x_{j2})^2 + (x_{i3} - x_{j3})^2} \tag{2.2}$$

In Eq. (2.1), $(\sigma/r)^{12}$ and $(\sigma/r)^{6}$ represent the repulsive and attractive interactions, respectively. The L-J potential curve is illustrated in Fig. 2.2, in which the attraction and repulsion regions are shown. In the repulsion region, the repulsion of the atoms rapidly increases as the distance between two atoms decreases. In the attraction region, as the distance between two

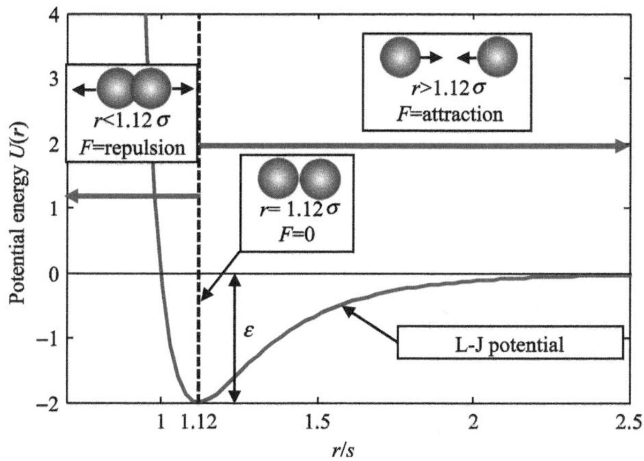

Fig. 2.2 L-J potential curve.

atoms increases towards a certain further separation, the attraction gradually drops to zero. When two atoms reach an equilibration distance ($r = 1.12\sigma$), their minimum bonding potential energy is reached. At this point, the interaction force between the atoms is equal to zero.

Having specified the potential energy function, the interaction force that the jth atom exerts on the ith atom is

$$F_{ij} = -\nabla U(r_{ij}) = \frac{24\varepsilon}{\sigma_2}\left[2\left(\frac{\sigma}{r_{ij}}\right)^{14} - \left(\frac{\sigma}{r_{ij}}\right)^{8}\right]\mathbf{r}_{ij} \tag{2.3}$$

So the total interaction force exerted on the ith atom is simply given as

$$F_i = \sum_{\substack{j=1 \\ j \neq i}}^{N} F_{ij} \tag{2.4}$$

where N is the total number of atoms in an atomic system.

To study more complex molecules, a molecular dynamics method with geometric constraints has been proposed (Ryckaert et al., 1977), in which a combination of geometrical constraints and internal motion of atoms is considered. In polyatomic molecules, the highest-frequency internal vibrations are usually decoupled from rotational and translational motions. Thus, a certain number of rigid bonds are introduced in the skeleton of the molecules. Consider the case in which the structure of a molecule is subject to one or more geometries. A constraint needs to be introduced to fix the distance between any two atoms with covalent bonds, and the mode can be expressed as

$$|\boldsymbol{x}_i - \boldsymbol{x}_i|^2 = b_{ij}^2 \tag{2.5}$$

where b_{ij} is the fixed bond length between the ith and jth atoms. Suppose that there are a total of l constraints influencing a particular molecule, and if the kth constraint for a bond works between the ikth and jkth atoms, then the kth constraint is

$$\theta_k = \left|\boldsymbol{x}_{ik} - \boldsymbol{x}_{jk}\right|^2 - b_{ij}^2 = 0, \quad k = 1, 2, \cdots, l \tag{2.6}$$

Hence, the constraint force G_i from the stretch of a covalent bond between two atoms acted on the ith atom can be written as

$$G_i = -\sum_{k=1}^{l} \lambda_k \nabla_i \theta_k = -2\sum_{k=1}^{l} \lambda_k \left(\boldsymbol{x}_{ik} - \boldsymbol{x}_{jk}\right) \tag{2.7}$$

where λ_k is the Lagrangian multiplier associated with θ_k. Hence, the motion equation of atoms with the constraint can be modified as

$$F_i + G_i = m_i a_i \tag{2.8}$$

For Eq. (2.8), the forces exerted on the atoms include not only all non-constraint interaction forces among molecules, but also the constraint force(s) within each molecule, thus embodying the essence of atomic motion.

In summary, basic molecular dynamics describes the movement principles of atoms, including the characteristics of the potential function, the motion mode of atoms with a nonconstraint interaction force, and a geometric constraint force. Despite the simplicity of the analytical model, the physics-based study of molecular dynamics can be used to determine thermodynamic properties of the system, and indeed presents opportunities for many theoretical studies and practical applications (Tuckerman et al., 1993; Pishkenari et al., 2016; Kilymis et al., 2016; Hasheminasab and Mehdipour, 2016).

2.3 Atom search optimization

In this section, a novel optimization algorithm termed atom search optimization (ASO), inspired by molecular dynamics, is introduced (Zhao et al.,

2019). In ASO, the position of each atom within the search space represents a solution measured by its mass, with a better solution indicating a heavier mass, and vice versa. All atoms in the population will attract or repel each other according to the distance among them, encouraging the lighter atoms to move towards the heavier ones. Heavier atoms have smaller acceleration, which makes them seek intensively for better solutions in local spaces. Lighter atoms have greater acceleration, which makes them search extensively to find new promising regions in the entire search space.

The general unconstrained optimization problems can be defined as

$$\text{Minimize}\, f(\boldsymbol{x}), \quad \boldsymbol{x} = \left(x^1, \cdots, x^d\right) \tag{2.9}$$

for

$$Lb \leq \boldsymbol{x} \leq Ub, Lb = \left[lb^1, \cdots, lb^d\right], Ub = \left[ub^1, \cdots, ub^d\right] \tag{2.10}$$

where $x^k(i = 1, \cdots, N)$ is the kth component of the search space, lb^k and ub^k are the kth component of the lower and upper limits, respectively, and d is the dimension of the search space.

In order to solve this unconstrained optimization, suppose an atom population with N atoms. The position of the ith atom is expressed as

$$\boldsymbol{x}_i = \left[x_i^1, \cdots, x_i^d\right], \quad i = 1, \cdots, N \tag{2.11}$$

where $x_i^k(k = 1, \cdots, d)$ is the kth position component of the ith atom in the dth dimension. In the initial iterations of ASO, each atom interacts with others by the attraction or the repulsion among them, and the repulsion can avoid the overconcentration of atoms and the premature convergence of the algorithm, thus enhancing the exploration ability in the entire search space. As iterations pass, the repulsion gradually weakens and the attraction gradually strengthens, which signifies that the exploration decreases and the exploitation increases. In the final iterations, each atom interacts with others just by the attraction, which ensures that the algorithm has a good exploitation capability.

2.3.1 Mathematical representation of interaction force

The interaction force resulting from the L-J potential is the priming power of atomic motion. The interaction force acted on the ith atom from the jth atom in the dth dimension at the tth iteration in Eq. (2.3) can be rewritten as

$$F_{ij}(t) = \frac{24\varepsilon(t)}{\sigma(t)}\left\{2\left[\frac{\sigma(t)}{r_{ij}(t)}\right]^{13} - \left[\frac{\sigma(t)}{r_{ij}(t)}\right]^{7}\right\}\frac{r_{ij}(t)}{\boldsymbol{r}_{ij}^d(t)} \tag{2.12}$$

and

$$F'_{ij}(t) = \frac{24\varepsilon(t)}{\sigma(t)}\left\{2\left[\frac{\sigma(t)}{r_{ij}(t)}\right]^{13} - \left[\frac{\sigma(t)}{r_{ij}(t)}\right]^{7}\right\} \tag{2.13}$$

The force curve of atoms is shown in Fig. 2.3. As shown, the atoms keep a relative distance, varying in a certain range all the time from the repulsion or attraction, and the change amplitude of the repulsion relative to the equilibration distance ($r = 1.12\sigma$) is much greater than that of the attraction. However, this model cannot be used directly to handle optimization problems, mainly because ASO needs to obtain more positive attraction and less negative repulsion as iterations increase, as shown in Fig. 2.3. Eq. (2.13) cannot satisfy this point. So, a revised version of this equation is developed, as follows, to solve the optimization problems

$$F'_{ij}(t) = -\eta(t)\left\{2\left[h_{ij}(t)\right]^{13} - \left[h_{ij}(t)\right]^{7}\right\} \tag{2.14}$$

where $\eta(t)$ is the depth function to adjust the repulsion region or attraction region, which can be defined as

$$\eta(t) = \alpha\left(1 - \frac{t-1}{T}\right)^{3} e^{-\frac{20t}{T}} \tag{2.15}$$

where α is the depth weight and T is the maximum number of iterations. The function behaviors of F', with different η corresponding to h ranging from 0.9 to 2, are illustrated in Fig. 2.4. From the figure, the repulsion occurs when h ranges from 0.9 to 1.12, the attraction occurs when h is between 1.12

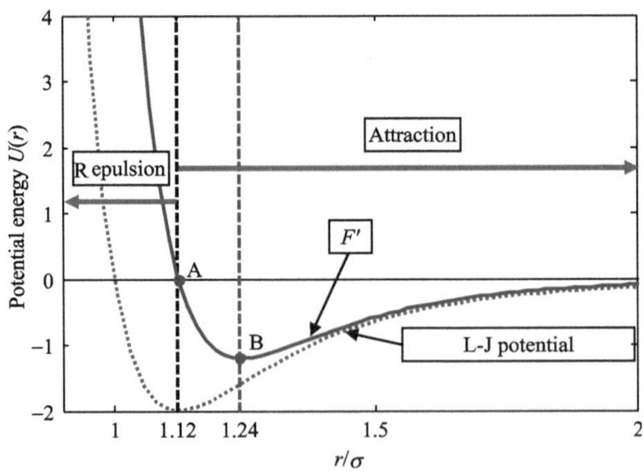

Fig. 2.3 Force curve of atoms.

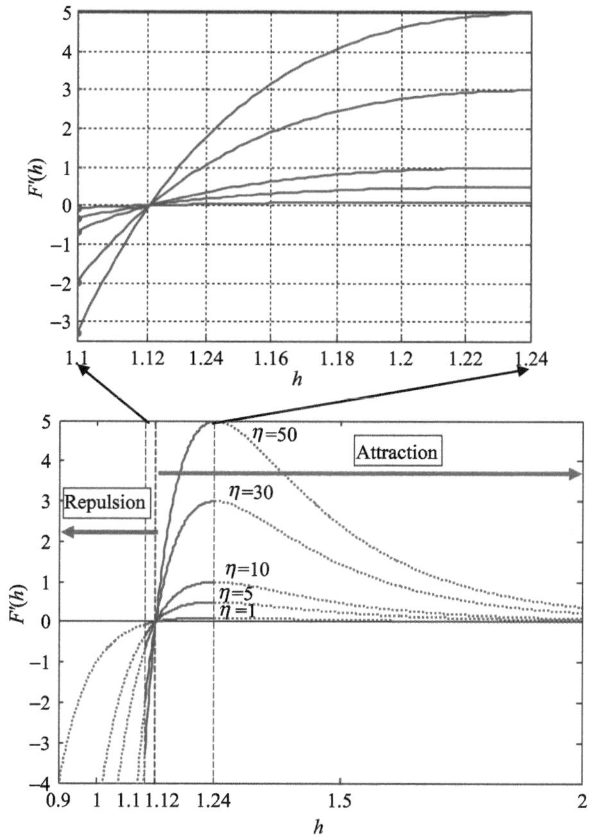

Fig. 2.4 Function behaviors of F' with different values of η.

and 2, and the equilibration occurs when $h = 1.12$. The attraction gradually increases with the increase of h from the equilibration ($h = 1.12$), reaches a maximum ($h = 1.24$) and then begins to decrease. The attraction approximately equals to 0 when h is greater than or equals to 2.

Therefore, in ASO, to improve the exploration, a lower limit of the repulsion with a smaller function value is set to $h = 1.1$ and an upper limit of attraction with a larger function value is set to $h = 1.24$. Therefore, h is defined as

$$h_{ij}(t) = \begin{cases} h_{\min}, & \dfrac{r_{ij}(t)}{\sigma(t)} < h_{\min} \\ \dfrac{r_{ij}(t)}{\sigma(t)}, & h_{\min} \le \dfrac{r_{ij}(t)}{\sigma(t)} \le h_{\max} \\ h_{\max}, & \dfrac{r_{ij}(t)}{\sigma(t)} > h_{\max} \end{cases} \tag{2.16}$$

where $h_{\min}$ and $h_{\max}$ are the lower and the upper limits of h, respectively, and the length scale $\sigma(t)$ is defined as

$$\sigma(t) = \left\| x_{ij}(t), \frac{\sum_{j \in K_{\text{best}}} x_{ij}(t)}{K(t)} \right\|_2 \tag{2.17}$$

and

$$\begin{cases} h_{\min} = g_0 + g(t) \\ h_{\max} = u \end{cases} \tag{2.18}$$

where K_{best}, which is a subset of an atom population, is made up of the first K atoms with the best function fitness values. As a drift factor, g can make the algorithm drift from the exploration to the exploitation and is given as

$$g(t) = 0.1 \times \sin\left(\frac{\pi}{2} \times \frac{t}{T}\right) \tag{2.19}$$

Then the sum of components with random weights acted on the ith atom from the other atoms can be considered a total force, which is expressed as

$$F_i^d(t) = \sum_{j \in K_{\text{best}}} \text{rand}_j F_{ij}^d(t) \tag{2.20}$$

where rand_j is a random number in [0, 1].

2.3.2 Mathematical representation of geometric constraint

The geometric constraint in molecular dynamics plays an important role in atomic motion. For simplicity, suppose each atom in ASO has a covalence bond with the best atom. Thus, each atom is acted on by a constraint force from the best atom, so the constraint of the ith atom can be rewritten as

$$\theta_i(t) = \left[|\boldsymbol{x}_i(t) - \boldsymbol{x}_{\text{best}}(t)|^2 - b_{i,\text{best}}^2 \right] \tag{2.21}$$

where $\boldsymbol{x}_{\text{best}}(t)$ is the position of the best atom at the tth iteration, and $b_{i,\text{best}}$ is a fixed bond length between the ith atom and the best atom. Hence the constraint force can be obtained as

$$G_i^d(t) = -\lambda(t) \nabla \theta_i^d(t) = -2\lambda(t) \left[x_i^d(t) - x_{\text{best}}^d(t) \right] \tag{2.22}$$

where $\lambda(t)$ is the Lagrangian multiplier. Then, making the substitution of $2\lambda \to \lambda$, the constraint force can be redefined as

$$G_i^d(t) = \lambda(t) \left[x_{\text{best}}^d(t) - x_i^d(t) \right] \tag{2.23}$$

The Lagrangian multiplier is defined as

$$\lambda(t) = \beta e^{-\frac{20t}{T}} \tag{2.24}$$

where β is the multiplier weight.

2.3.3 Mathematical representation of atomic motion

With the interaction force and the geometric constraint, the acceleration of the ith atom at time t can be written as

$$a_i^d(t) = \frac{F_i^d(t)}{m_i^d(t)} + \frac{G_i^d(t)}{m_i^d(t)} = -\alpha\left(1 - \frac{t-1}{T}\right)^3 e^{-\frac{20t}{T}} \sum_{j \in K_{best}} \frac{\text{rand}_j\left[2 \times \left(h_{ij}(t)\right)^{13} - \left(h_{ij}\right)^7\right]}{m_i(t)} \frac{\left[x_i^d(t) - x_i^d(t)\right]}{\left\|\boldsymbol{x}_i(t), \boldsymbol{x}_j(t)\right\|_2} + \beta e^{-\frac{20t}{T}} \frac{x_{best}^d(t) - x_i^d(t)}{m_i(t)} \tag{2.25}$$

where $m_i(t)$ is the mass of the ith atom at the tth iteration, which can be measured at the simplest level by its function fitness value. The mass of the ith atom can be calculated as

$$M_i(t) = e^{-\frac{Fit_i(t) - Fit_{best}(t)}{Fit_{worst}(t) - Fit_{best}(t)}} \tag{2.26}$$

$$m_i(t) = \frac{M_i(t)}{\sum_{j=1}^{N} M_j(t)} \tag{2.27}$$

where $Fit_{best}(t)$ and $Fit_{worst}(t)$ are the atoms with the minimum fitness value and the maximum fitness value at the tth iteration, respectively. $Fit_i(t)$ is the function fitness value of the ith atom at the tth iteration. $Fit_{best}(t)$ and $Fit_{worst}(t)$ are expressed as

$$Fit_{best}(t) = \min_{i \in \{1, 2, \cdots, N\}} Fit_i(t) \tag{2.28}$$

$$Fit_{worst}(t) = \max_{i \in \{1, 2, \cdots, N\}} Fit_i(t) \tag{2.29}$$

To simplify the algorithm, the position and velocity of the ith atom at the $(t + 1)$th iteration can be denoted as follows:

$$v_i^d(t+1) = \text{rand}_i^d v_i^d(t) + a_i^d(t) \tag{2.30}$$

$$x_i^d(t+1) = x_i^d(t) + v_i^d(t+1) \tag{2.31}$$

In the ASO algorithm, to enhance the exploration in the first stage of iterations, each atom needs to interact with as many atoms with better fitness values as its K neighbors. To enhance the exploitation in the final stage of iterations, the atoms need to interact with as few atoms with better fitness values as its K neighbors. So, as a function of time, K_{best} gradually decreases with the lapse of iterations. K_{best} can be calculated as

$$K_{\text{best}}(t) = N - (N-2) \times \sqrt{\frac{t}{T}} \tag{2.32}$$

The forces of an atom population are shown in Fig. 2.5. As shown in the figure, A_1, A_2, A_3, and A_4 compose the K_{best} neighbors. A_5, A_6, and A_7 attract or repel each atom in the K_{best} neighbors and A_1, A_2, A_3, and A_4 attract or repel each other. Each atom in the population except for A_1 - (x_{best}) has a constraint force from the best atom A_1.

A simulation is conducted to examine how atoms move with this mathematical model. The swarm motion of five atoms around a target in a three-dimensional space is illustrated in Fig. 2.6, in which five balls represent five different atoms, and the ball in the central of the cube represents the desired target that every atom wants to reach. Initially, the positions of the five atoms are randomly generated in the search space. With the lapse of time t, all the atoms gradually approach the target using the mathematical mode and form a swarm. Finally, all the atoms converge to the target. Additionally, it can be found that, although one of the atoms is far away from the swarm when $t = 20$, the other atoms also pull it back by the attraction in the subsequent

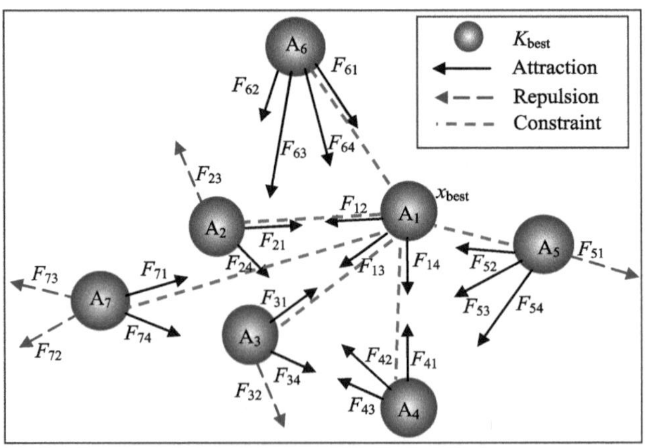

Fig. 2.5 Forces of an atom system with K_{best} for $K = 5$.

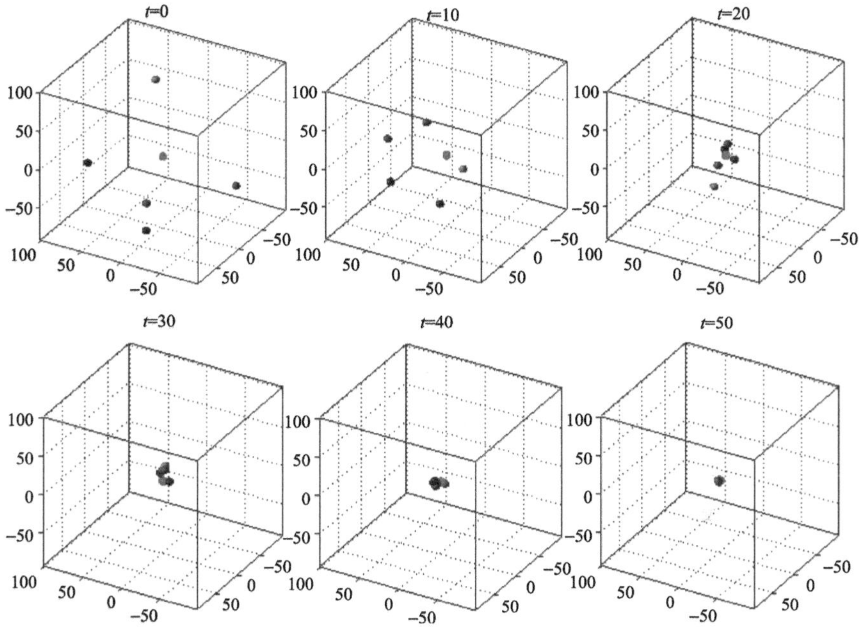

Fig. 2.6 Swarm motion of five atoms around a target in a three-dimensional space.

iterations, and all the atoms do not become too concentrated because of the repulsion. The motion histories of the five atoms during 50 iterations are illustrated in Fig. 2.7. It is apparent that the atoms grow denser when they are closer to the target, and the distribution of atoms in the search space is sufficient to demonstrate that the model proposed can achieve the transition from the exploration for the entire search space to the exploration for a focused region. It is obvious that this search characteristic can be extended to a n-D space.

2.3.4 Framework of the ASO algorithm

ASO starts the optimization by generating a set of random solutions. The atoms update their positions and velocities in each iteration, and the position of the best atom so far is also updated in each iteration. In addition, the acceleration of atoms comes from two parts. One is the interaction force caused by the L-J potential, which actually is the vector sum of the attraction and the repulsion exerted from other atoms. Another is the constraint force caused by the bond-length potential, which is the weighted position difference between each atom and the best atom. All the updating and the calculation are performed interactively until the stop criterion is satisfied. Finally,

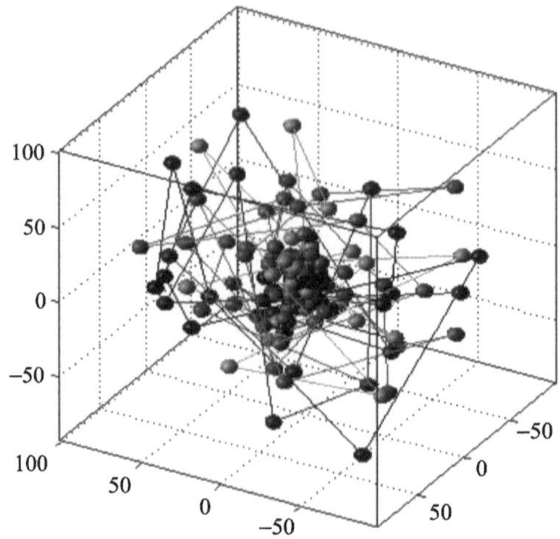

Fig. 2.7 Motion histories of five atoms during 50 iterations.

the position and the fitness value of the best atom are returned as an approximation to the global optimum. The pseudo code of ASO algorithm is provided in Fig. 2.8.

2.4 Experimental results

2.4.1 Benchmark functions

To test the performance of the ASO algorithm extensively, it is employed to solve 37 well-known benchmark functions. These functions can be divided into five types: unimodal, multimodal, low-dimensional, hybrid, and composite functions, by which the performance of many different optimization algorithms can be efficiently measured. These benchmark functions (Digalakis and Margaritis, 2011; Liang et al., 2014; Zhao and Wang, 2016) are summarized in Appendix Tables A.1–A.4. Functions f_1–f_7 are unimodal functions and each has only one global optimum and no local optimum, so the convergence rate and exploitation of algorithms can be verified. Functions f_8–f_{13} are multimodal functions with a considerable number of local optimum. Functions f_{14}–f_{23} are low-dimensional functions, each of which has fewer local optimum. These multimodal and low-dimensional functions with local optimum are highly suitable for test avoidance of local optima and exploration capacity of the algorithms. Functions f_{24}–f_{29} are hybrid functions, in which their variables are randomly separated into different subdivisions. These subdivisions are replaced by using either

Randomly initialize a set of atoms X (solutions) and their velocity v, and Fit_{best}=Inf.
While the stop criterion is not satisfied do
  For each atom X_i do
    Calculate the fitness value Fit_i;
    If $Fit_i < Fit_{best}$ then
      $Fit_{best} = Fit_i$;
      $X_{best} = X_i$;
    End If.
    Calculate the mass using Eqs. (2.26) and (2.27);
    Determine its K neighbors using Eq. (2.32);
    Calculate the intraction force F_i and the constraint force G_i using Eq. (2.20) and (2.23), respectively;
    Calculate the acceleration using Eq. (2.25);
    Update the velocity using Eq. (2.30);
    Update the position using Eq. (2.31);
  End For.
End While.
Find the best solution so far X_{best}.

Fig. 2.8 Pseudo code of ASO algorithm.

unimodal or multimodal functions. Additionally, functions f_{30}–f_{37} are composition functions, their variables are also randomly separated into different subdivisions. These subdivisions are constructed by using the basic and hybrid functions. Each composition function has different properties for different subdivisions. The global optimum is shifted from a specific position to a random position before each iteration, and occasionally the optimum is relocated on the boundary of the search space (Mirjalili and Lewis, 2016). Hybrid and composition test functions are more complex and challenging than the basic unimodal and multimodal test functions, thus resulting in more difficult optimization. Accordingly, they are especially suitable for testing the potential performance of the algorithms for solving real-world problems. A detailed description of these functions is available in Rashedi et al. (2009).

2.4.2 Experimental setup

For these test functions, the proposed algorithm is compared with five stochastic optimization algorithms, including three classic and popular algorithms, PSO, GA, and SA, and two recently proposed algorithms, GSA and WDO. Although a number of variants based on these algorithms have been developed, the comparisons of standard versions can be used to interpret the results of larger groups (Zhao et al., 2019).

PSO (Kennedy and Eberhart, 1995) mimics the behaviors of flocks of birds in the sky. It updates velocities and positions of a population by using social group learning and individual learning to seek expected goals. PSO has a good local search ability.

GA (Holland, 1975), an evolutionary algorithm, is inspired by the biology evolutionary theory. It uses mutation, crossover, and selection operations to generate high-quality solutions. GA tends to be excellent for finding good global solutions.

SA (Kirkpatrick et al., 1983) takes inspiration from annealing in metallurgy by heating material and cooling it at a certain rate. It is a probabilistic algorithm for seeking the global optimum in the search space and in a fixed amount of time.

GSA (Rashedi et al., 2009), based on the law of gravity, makes agents interact according to the law of motion and is a competitive search algorithm. This interaction can generate an attractive force that facilitates all agents globally moving to the agent with the heavier mass.

WDO (Bayraktar et al., 2013) is based on the earth's atmosphere motion wherein each small air parcel moves following Newton's second law. It updates the velocity and position of each parcel by the sum of a gradient force, Coriolis force, gravitational force, and friction force to approximate the global optimum. WDO has a good global search ability.

It should be noted that, for each benchmark function, it is time-consuming and difficult for each algorithm to search for a set of appropriate parameters to enhance its optimization performance. Hence, for the sake of fairness, a set of fixed parameters were adopted for each algorithm to evaluate the entire performance of all the test functions. The initial parameters used in all algorithms are provided in Table 2.1. In this experiment, the size of the population and the maximum number of iterations are set to 50 and 1000, respectively, for all the algorithms. In addition, every algorithm runs 50 times for each function, and the results are based on the average performance of the 50 runs.

2.4.3 Results and discussion

2.4.3.1 Qualitative results of ASO

First, five different functions are randomly chosen from five types of benchmark functions and simulated to see the effects of the proposed ASO algorithm qualitatively. The functions chosen are f_5, f_9, f_{15}, f_{25}, and f_{32} with

Table 2.1 Parameter settings for all algorithms.

Algorithm	parameter	Value
ASO	Depth weight	50
	Multiplier weight	0.2
WDO	RT coefficient	3
	Gravitational constant	0.2
	Constant in the update equations	0.4
	Coriolis effect	0.4
	Maximum allowable speed	0.3
SA	Initial temperature	0.1
	Temperature reduction rate	0.98
	Mutation rate	0.5
PSO	Cognitive constant	2
	Social constant	2
	Inertia constant	Linearly decrease from 0.8 to 0.2
GSA	Initial gravitational constant	100
	Decreasing coefficient	20
GA	Selection	Roulette wheel
	Crossover	0.8
	Mutation	0.4

$D = 2$, whose shapes are presented in Figs. 2.9–2.13, respectively. ASO is used for their minimization using four atoms during 150 iterations. The diversity of the benchmark functions makes for comprehensive and easy observation of the algorithm performance from different views (Zhao et al., 2019).

Four qualitative metrics used to describe the performance of ASO include search history, trajectory curve, convergence curve, and interaction force. All history positions of the four atoms during the iterations are clearly depicted in the figures, and the distribution density of atoms in the search space can indicate how ASO performs the explorative and exploitative search. Apparently, the low distribution density indicates the explorative search for the global space, and the high distribution density indicates the exploitative search for the local space. It is evident that the distribution of atoms is sparse in the region further from the global optimum and is dense in the region closer to the global optimum.

The trajectory curve is one of the most important metrics that can effectively depict exploration and exploitation of the algorithm. It can trace

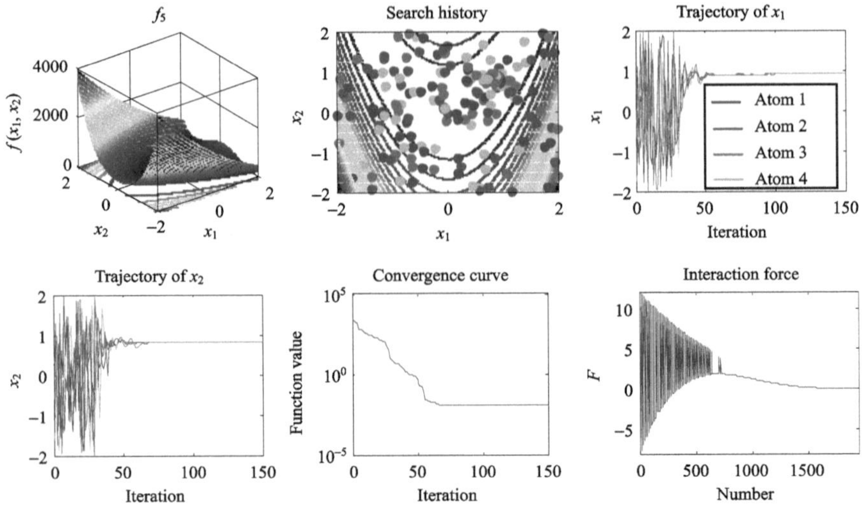

Fig. 2.9 Qualitative results of f_5.

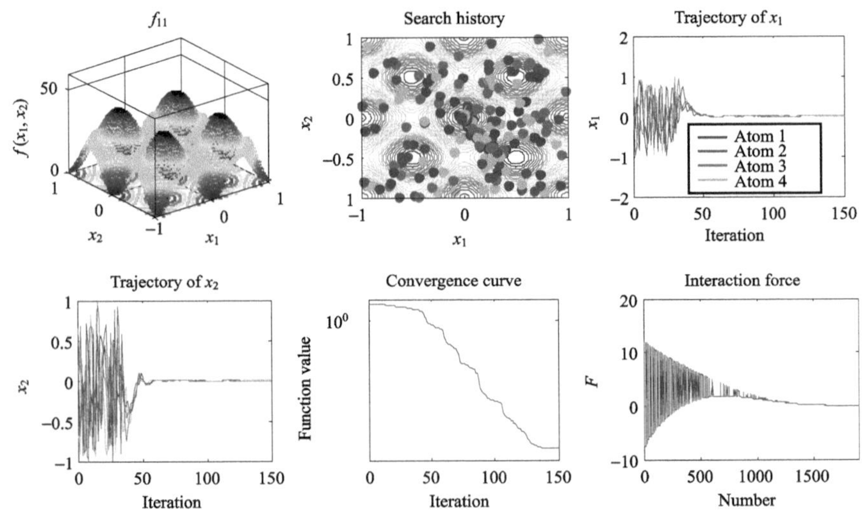

Fig. 2.10 Qualitative results of f_{11}.

positions of all atoms in different dimensions during iterations. Figs. 2.9–2.13 illustrate the trajectory curves when the five functions are optimized using the ASO algorithm. In the figures, the trajectory curves of four atoms in the first and the second dimensions are depicted. All trajectory curves show frequent large-scale fluctuations in the early iterations. With the lapse of

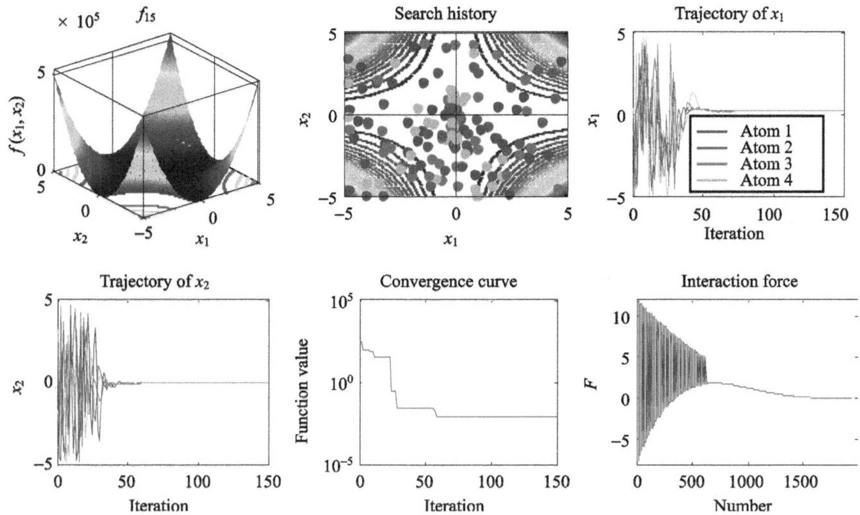

Fig. 2.11 Qualitative results of f_{15}.

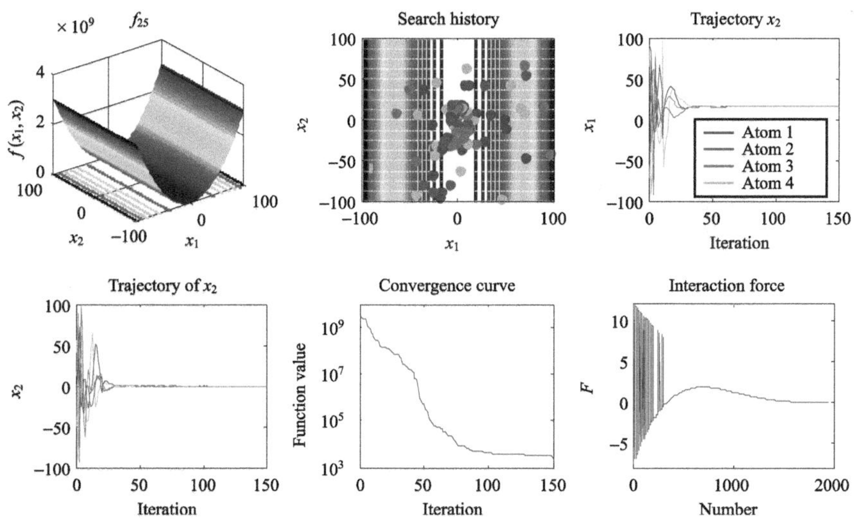

Fig. 2.12 Qualitative results of f_{25}.

iterations, such variation decreases in amplitude or frequency, and the positions of atoms become monotonous and gradually tend to stabilize to the global optimum in the later iterations. Evidently, the large-scale fluctuations in the former iterations indicate the explorative search for the global space, and the small-scale fluctuations in the latter iterations indicate the exploitative search for the local space. These two behaviors are consistent with what

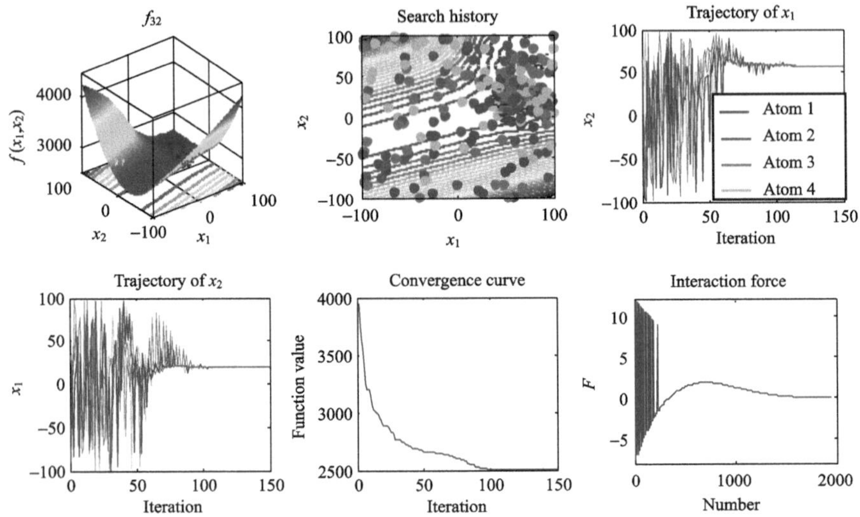

Fig. 2.13 Qualitative results of f_{32}.

is shown in the search histories. It can also be observed that ASO performs the exploration before the exploitation in the trajectory curves. Therefore, the trajectory curves together with the search histories reveal how well ASO conducts the explorative and exploitative search.

For all optimization algorithms, the final goal is to find the approximation to the global optimum quickly and precisely, so being able to show this behavior is of great importance. But the trajectory curves and search histories cannot help. The convergence curve is the most often used qualitative metric in evaluating the convergence performance of the algorithms. It is apparent from the figures that the convergence curves can clearly depict the convergence rate and the approximation to the global optimum. As shown in the figures, the convergence curve of function f_5 is very smooth and drops rapidly, demonstrating that the exploitation contributes more to ASO than the exploration. In contrast, for functions f_{11}, f_{15}, f_{25}, and f_{32}, their convergence curves are very rough and drop slowly, which indicates that the exploration contributes more to ASO than the exploitation. Finally, their convergence curves can all accurately approximate the global optimum in the final iterations.

A large number of real-world problems can be considered multimodal functions with a considerable number of local optima (Cuevas and González, 2013). When solving these problems, the exploration must avoid the local optima and discover a promising region with the global optimum.

In Figs. 2.9–2.13, the interaction forces between every two atoms during the whole iterations are provided. The interaction forces are shown as either the repulsion (negative) or the attraction (positive) in the former stage and only as the attraction in the latter stage. Additionally, the interaction forces gradually decrease with the lapse of iterations. The attraction in the latter stage can prevent the atom swarm from prematurely concentrating in a local region, which can contribute significantly to the exploration of ASO.

Based on the above results and discussion, the four qualitative metrics can show how well the exploration and exploitation perform, the convergence performance, and the accuracy of the final approximation to the global optimum. The results indicate that ASO establishes a proper balance between the exploration and exploitation. In addition, the search performance for the global optimum is also satisfactory.

2.4.3.2 Convergence preference of the algorithm

Although the qualitative metric results reveal that ASO has a good optimization performance, they cannot indicate how well the proposed ASO performs for all benchmark functions. In this subsection, ASO is evaluated using some performance evaluation indexes, which are compared to those of the other algorithms. To obtain statistical solutions and better evaluate the algorithms, there are three performance evaluation indexes used to quantitatively compare all the algorithms: the average, the standard deviation, and minimum of the best-so-far solution. Among all the 50 runs, the lower the average of the best-so-far solution, the greater the algorithm's ability to avoid local optima and approximate a global optimum. Also, the lower the standard deviation of the best-so-far solution, the closer the solution is to the average. Further, the more stable the algorithm is, the better the minimum of the best-so-far solution, and the more accurate the approximation is to the global optimum. The average of the best-so-far solution among the comparative algorithms on each function is shown in bold in the tables below (Zhao et al., 2019).

Tables 2.2–2.6 show the comparisons of optimization results obtained for different types of functions given in Appendix Tables A.1–A.4, and the convergence processes of all comparative algorithms are shown in Figs. 2.14–2.18. As described in Wolpert and Macready (1997), there is no optimization algorithm that can best perform for all different types of problems, hence the performance evaluation of the proposed algorithm should come from different perspectives by using multifarious benchmark functions.

Table 2.2 Comparisons of results for unimodal functions.

Function	Index	WDO	SA	PSO	GSA	GA	ASO
f_1	Mean	**0**	2.04E−13	0.000146	2.11E−17	0.010212	2.68E−21
	Std	0	6.14E−14	0.000119	6.67E−18	0.005073	3.65E−21
	Best	0	7.76E−14	9.56E−06	1.06E−17	0.002975	3.52E−22
f_2	Mean	**0**	9.73102	0.000198	2.37E−08	0.020789	3.33E−10
	Std	0	7.34169	0.000188	2.60E−09	0.005434	1.89E−10
	Best	0	0.003379	5.45E−05	1.71E−08	0.012651	5.34E−11
f_3	Mean	**6.73E−29**	4554.414	2659.282	251.2258	1115.564	197.5452
	Std	3.69E−28	2329.664	1194.893	41.96643	493.2294	79.7024
	Best	0	1527.615	760.8198	179.2156	401.8981	23.1228
f_4	Mean	**0**	6.399475	18.26166	3.35E−09	0.994831	3.24E−09
	Std	0	3.284424	4.247802	7.53E−09	0.354854	6.14E−09
	Best	0	1.338689	10.65944	2.28E−09	0.547858	2.13E−10
f_5	Mean	28.19528	1059.101	134.3922	28.1473	97.11795	**24.8388**
	Std	0.168679	2050.492	128.5852	11.29836	128.7488	0.515853
	Best	27.95513	23.1826	26.23857	25.75286	9.696092	16.58185
f_6	Mean	**0**	0.566667	0.133333	**0**	**0**	**0**
	Std	0	0.727932	0.345746	0	0	0
	Best	0	0	0	0	0	0
f_7	Mean	**6.10E−05**	0.123989	0.066316	0.02076	0.0505	0.035641
	Std	4.49E−05	0.039739	0.019519	0.007716	0.021781	0.019498
	Best	1.43E−06	0.062475	0.029819	0.007006	0.0132	0.03615

Table 2.3 Comparisons of results for multimodal functions.

Function	Index	WDO	SA	PSO	GSA	GA	ASO
f_8	Mean	−5840.86	**−9289.96**	−5197.85	−2653.67	−6795.16	−7428.17
	Std	857.0229	402.1469	529.3583	341.6006	632.7758	422.3977
	Best	−7594.51	−10,055.4	−6787.39	−3488.44	−8147.59	−5561.44
f_9	Mean	57.69854	54.39099	29.31833	15.05704	12.52966	**0**
	Std	21.2409	13.82194	6.879953	4.439862	2.915823	0
	Best	15.56306	26.86388	17.34745	7.959667	6.094298	0
f_{10}	Mean	**8.88E−16**	0.343794	0.007432	3.69E−09	0.021188	3.00E−11
	Std	0	0.447427	0.014159	3.96E−10	0.004548	2.15E−11
	Best	8.88E−16	8.29E−08	0.000549	2.96E−09	0.010696	1.13E−11
f_{11}	Mean	0.009891	0.012439	0.022795	4.472721	0.018359	**0**
	Std	0.021887	0.010314	0.02737	2.048563	0.009772	0
	Best	0	3.16E−06	5.88E−05	1.944629	0.006121	0
f_{12}	Mean	0.03139	0.698388	0.337564	0.020732	2.80E−05	**4.51E−23**
	Std	0.087872	0.647321	0.429656	0.057102	1.89E−05	1.88E−23
	Best	0.000355	0.046437	2.54E−05	7.00E−20	7.49E−06	8.69E−25
f_{13}	Mean	0.548891	0.514392	0.232059	0.00155	0.000452	**1.91E−23**
	Std	0.931434	1.324152	0.71351	0.003696	0.00069	3.12E−22
	Best	0.010295	2.77E−05	0.000487	1.31E−18	0.000119	2.31E−24

Table 2.4 Comparisons of results for low-dimensional functions.

Function	Index	WDO	SA	PSO	GSA	GA	ASO
f_{14}	Mean	5.311273	1.395219	**0.998004**	4.215728	3.92122	**0.998004**
	Std	5.651641	0.669666	3.41E−13	3.636458	2.697616	7.40E−17
	Best	0.998004	0.998004	0.998004	0.998021	0.998004	0.998004
f_{15}	Mean	0.001089	7.67E−04	**3.20E−04**	0.002059	0.001259	9.47E−04
	Std	0.003656	2.19E−04	3.64E−05	8.64E−04	0.001909	2.27E−04
	Best	3.08E−04	4.78E−04	3.07E−04	9.98E−04	3.25E−04	2.79E−04
f_{16}	Mean	−1.03162	**−1.03163**	**−1.03163**	**−1.03163**	**−1.03163**	**−1.03163**
	Std	1.00E−05	6.39E−16	6.78E−16	5.68E−16	7.56E−10	0
	Best	−1.03163	−1.03163	−1.03163	−1.03163	−1.03163	−1.03163
f_{17}	Mean	0.39791	**0.397887**	**0.397887**	**0.397887**	**0.397887**	**0.397887**
	Std	1.90E−05	0	6.09E−08	0	3.95E−08	0
	Best	0.397888	0.397887	0.397887	0.397887	0.397887	0.397887
f_{18}	Mean	3.000224	**3**	**3**	**3**	3.9	**3**
	Std	0.000385	1.28E−15	1.32E−15	1.99E−15	4.929503	1.51E−15
	Best	3	3	3	3	3	3
f_{19}	Mean	−3.86273	**−3.86278**	**−3.86278**	**−3.86278**	**−3.86278**	**−3.86278**
	Std	0.00011	2.71E−15	2.70E−15	2.46E−15	2.72E−09	2.68E−15
	Best	−3.86278	−3.86278	−3.86278	−3.86278	−3.86278	−3.86278
f_{20}	Mean	−3.26356	−3.27717	−3.2702	**−3.322**	−3.30614	**−3.322**
	Std	0.05942	0.056126	0.060244	1.36E−15	0.041107	1.12E−15
	Best	−3.32082	−3.322	−3.322	−3.322	−3.322	−3.322
f_{21}	Mean	−6.89682	**−9.61699**	−5.53835	−7.18959	−8.73973	−8.774464
	Std	3.606535	1.642171	2.55709	3.354687	2.904334	2.186718
	Best	−10.1532	−10.1532	−10.1532	−10.1532	−10.1532	−10.1532
f_{22}	Mean	−7.16466	**−10.4029**	−8.23528	**−10.4029**	−8.68393	**−10.4029**
	Std	3.580271	3.30E−16	3.17987	4.66E−16	3.176461	1.84E−15
	Best	−10.4029	−10.4029	−10.4029	−10.4029	−10.4029	−10.4029
f_{23}	Mean	−6.34454	**−10.5364**	−8.63826	**−10.5364**	−8.81243	**−10.5364**
	Std	4.002617	2.91E−15	3.015894	1.75E−15	3.212375	1.54E−15
	Best	−10.5363	−10.5364	−10.5364	−10.5364	−10.5364	−10.5364

Table 2.5 Comparisons of results for hybrid functions.

Function	Index	WDO	SA	PSO	GSA	GA	ASO
f_{24}	Mean	636,845.7	826,767.3	3,778,947	4,974,916	2,123,123	**452,502.66**
	Std	400,326.4	859,291.4	3,856,484	2,230,865	1,146,082	538,536.4
	Best	114,642.2	53,802.62	413,267.5	2,197,117	606,506.3	10,509.77
f_{25}	Mean	473,538.8	9897.176	1,892,268	**2360.049**	2922.662	2521.198
	Std	167,046.5	8871.518	9,041,101	402.0756	884.0251	1001.254
	Best	231,041.9	2319.467	1880.679	2051.498	1899.538	1973.163
f_{26}	Mean	1932.082	1912.644	1946.653	2048.729	1927.013	**1911.15**
	Std	30.64479	2.039958	33.06431	40.27749	24.00536	0.742535
	Best	1907.461	1907.34	1907.134	1929.327	1910.308	1909.686
f_{27}	Mean	**17,305.09**	57,189.46	28,438.85	146,057.1	39,156.38	21,999.82
	Std	6298.762	25,860.57	11,237.46	43,914.36	14,032.56	8835.043
	Best	4085.151	15,530.32	11,915.96	82,249.38	13,558.32	8941.514
f_{28}	Mean	222,605.1	200,164	793,350.6	1,330,821	460,277.3	**22,083.121**
	Std	106,098.9	193,452.9	761,226.8	1,065,693	504,584.7	16,323.2
	Best	57,142.06	33,146.05	**14,201.94**	346,135.4	67,981.71	16,885.179
f_{29}	Mean	2872.024	**2463.856**	2831.22	3267.385	2825.28	2756.953
	Std	183.498	116.4591	265.9089	293.3663	203.9491	204.2695
	Best	2431.908	2282.106	2370.504	2628.743	2348.308	2360.424

Table 2.6 Comparisons of results for composition functions.

Function	Index	WDO	SA	PSO	GSA	GA	ASO
f_{30}	Mean	2624.533	2632.702	2645.122	**2534.215**	2616.423	2616.1100
	Std	2.286212	10.07164	11.87371	63.42403	0.438999	1.941234
	Best	2620.749	2620.702	2628.129	2500	2615.828	2615.248
f_{31}	Mean	**2600**	2637.749	2632.222	2603.002	2630.226	2624.429
	Std	4.73E−05	5.807932	7.351382	6.04137	1.804637	4.75412
	Best	2600	2624.82	2625.289	2600.112	2627.5	2600.568
f_{32}	Mean	**2700**	2710.652	2722.703	2701.248	2713.621	2710.987
	Std	0	2.170978	7.350302	2.489568	2.764908	2.35979
	Best	2700	2707.026	2710.674	2700	2708.848	2700
f_{33}	Mean	2766.93	2706.882	2704.985	2793.516	2793.486	**2700.254**
	Std	4.169525	25.60925	18.07296	16.97263	25.30876	0.04997
	Best	2700.599	2700.073	2700.389	2747.814	2700.357	2700.133
f_{34}	Mean	3606.247	3206.937	3384.102	4372.258	3468.123	**3126.185**
	Std	309.7327	80.83431	208.4262	411.4156	345.4187	35.12669
	Best	3113.73	3110.855	3151.307	3054.283	3104.8	3102.133
f_{35}	Mean	7136.114	**3844.078**	7047.004	5169.593	6658.137	4795.1736
	Std	967.6085	162.4164	698.6452	884.0748	541.6948	286.4972
	Best	4984.829	3636.991	5780.443	3946.267	5431.646	3894.87
f_{36}	Mean	1,006,116	1,002,510	13,894,512	2,023,296	8380.851	**7366.303**
	Std	4,951,051	3,802,991	19,870,463	11,065,067	1290.275	1360.375
	Best	30,874.43	7012.566	5756.596	5100.102	5285.181	5024.042
f_{37}	Mean	62,783.62	29,698.5	148,713.6	195,076.3	13,699.9	**12,028.283**
	Std	22,308.01	11,159.28	174,091.8	88,263.52	3855.628	2022.0797
	Best	7525.18	12,019.36	16,603.94	8200.015	8526.658	7651.182

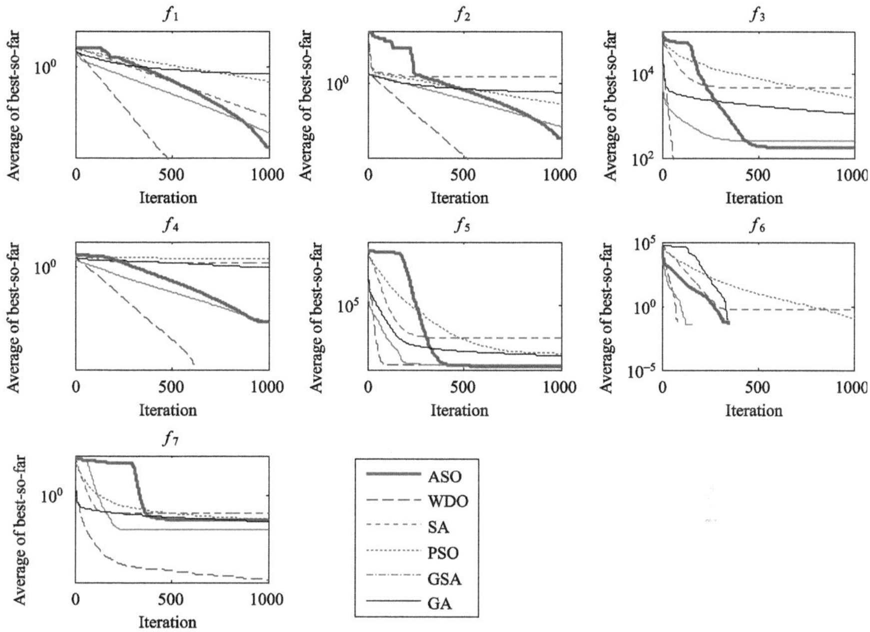

Fig. 2.14 Performance comparisons of algorithms for unimodal functions.

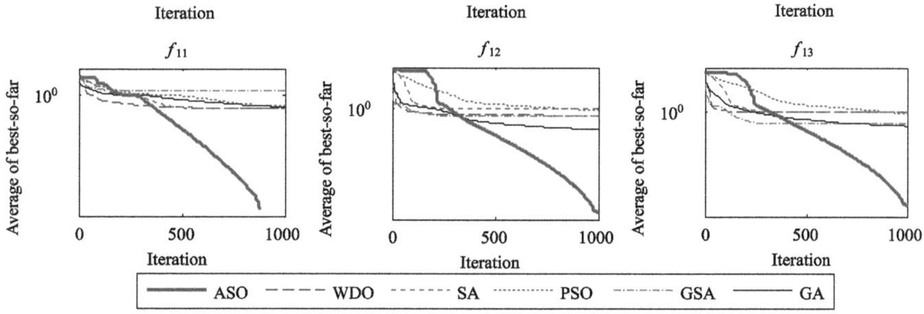

Fig. 2.15 Performance comparisons of algorithms for multimodal functions.

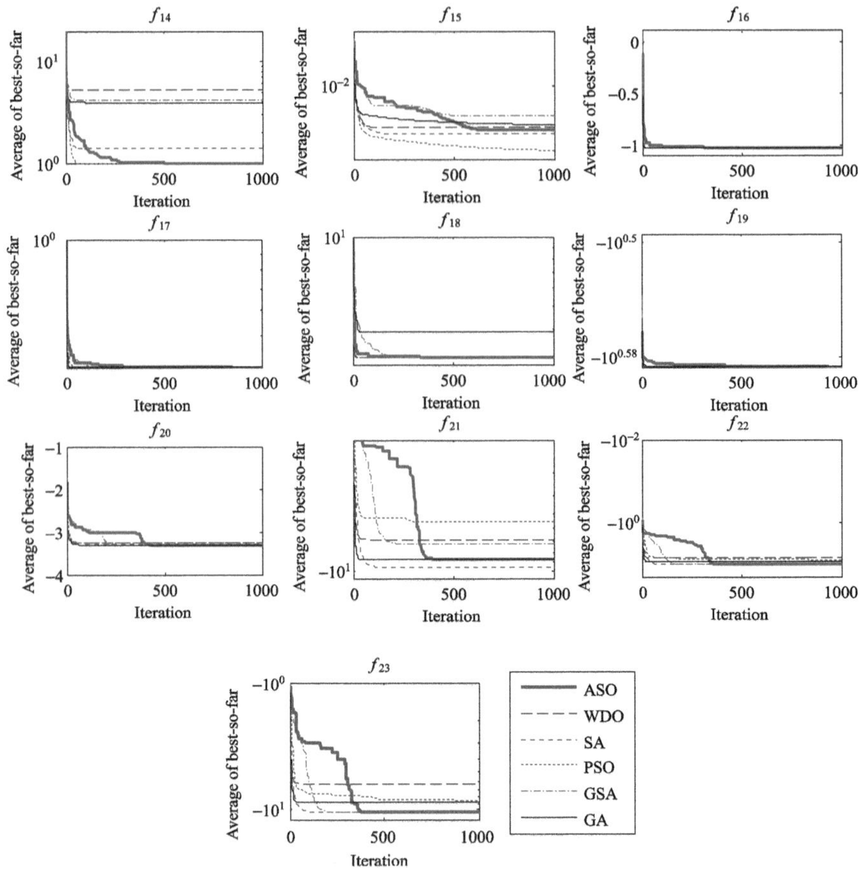

Fig. 2.16 Performance comparisons of algorithms for low-dimensional functions.

For the unimodal functions f_1–f_7, from Fig. 2.14 and Table 2.2, ASO performs better than the others, except for WDO on functions f_1–f_4, and performs the best for function f_5. Although ASO fails to achieve the best convergence performance for functions f_6 and f_7, it performs better than PSO, GA, and SA. Specifically, the convergence rate of ASO on some functions is slower than that of the other algorithms in the early iterations because of the repulsion among atoms, which makes them perform a global search in the entire search space to prevent the optimization from prematurity. In addition, WDO and GSA are also excellent and novel stochastic algorithms developed recently that have a certain advantage in terms of the convergence rate. Even so, ASO is still competitive in terms of the convergence rate for

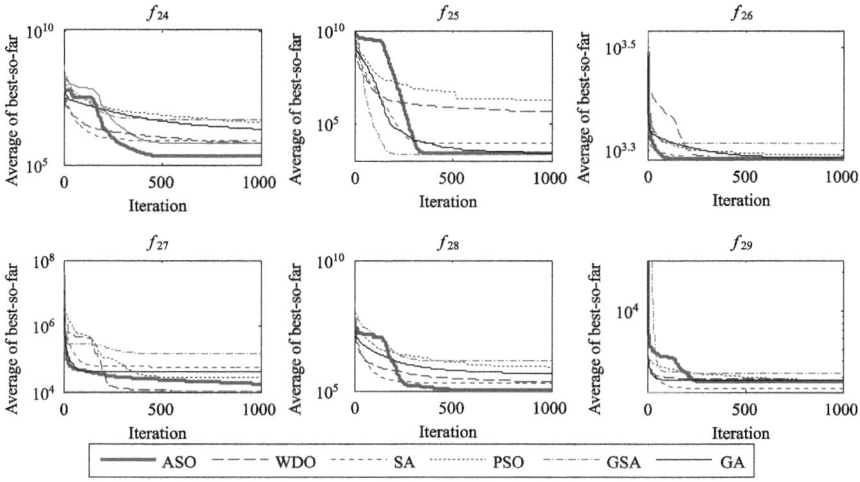

Fig. 2.17 Performance comparisons of algorithms for hybrid functions.

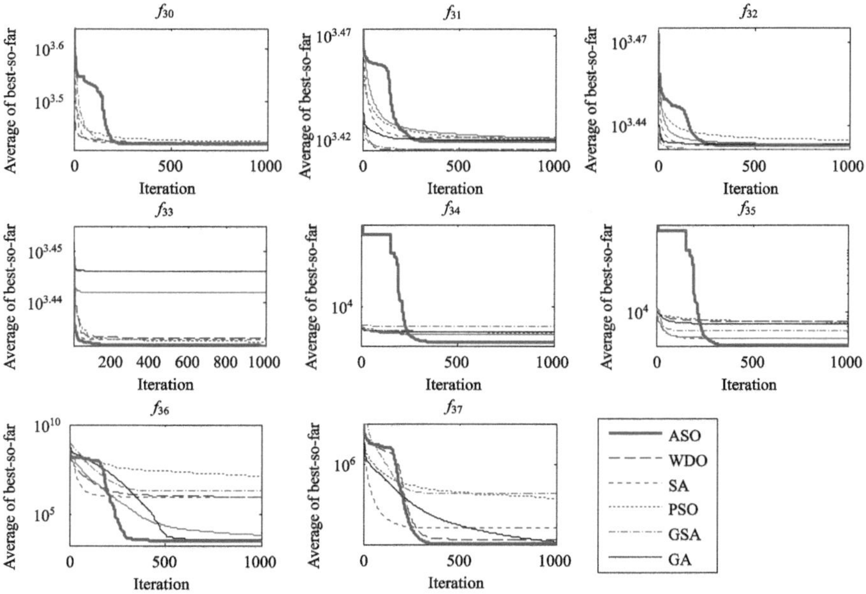

Fig. 2.18 Performance comparisons of algorithms for composition functions.

unimodal functions. The final solutions in Table 2.2 show that ASO achieves better results than the others on function f_5 and has the same results as WDO, GSA, and GA on function f_6. It ranks behind only WDO on functions f_1–f_4. Therefore, ASO can achieve a good precision for the unimodal functions.

For the multimodal functions f_8–f_{13}, from Fig. 2.15 and Table 2.3, ASO performs much better than the others, except for functions f_8 and f_{10}. ASO is inferior to SA on function f_8 and is outperformed by WDO on function f_{10}. However, WDO shows the worst results for functions f_9 and f_{13}, and GSA shows the worst results for functions f_8 and f_{11}. Although SA has the best results on function f_8, the algorithm shows worse results than the others on functions f_{10} and f_{12}, indicating that the three algorithms all fail to step out of the local optimum in some multimodal functions. For ASO, the repulsion can effectively avoid prematurity and greatly improve the exploration in the early stage. In the later stage, the stronger attraction force and the constraint force jointly contribute to the fine search for the global optimum.

For low-dimensional functions f_{14}–f_{23}, which are multimodal functions with low-dimensions and a few local optima. Fig. 2.16 and Table 2.4 show that all the algorithms provide nearly the same performance for functions f_{16}, f_{17}, and f_{19}. They also perform similarly well except for GA for function f_{18}, and ASO obtains the same good results as SA and GSA on functions f_{22} and f_{23}. ASO also performs the best for functions f_{14} and f_{20}, ranking behind only SA for function f_{21}. Even for function f_{15}, ASO performs better than WDO, GSA and GA. Obviously, ASO is highly competitive for these low-dimensional functions with local optima.

For hybrid functions f_{24}–f_{29}, Fig. 2.17 and Table 2.5 show that obviously, ASO performs the best on functions f_{24}, f_{26}, and f_{28}. For functions f_{25}, f_{27}, and f_{29}, ASO ranks only second to GSA, WDO and SA, respectively. In addition, GSA shows the worst convergence performance and results on functions f_{26}–f_{29}, and PSO performs the worst on functions f_{24} and f_{25}. It should be noted that the variables of the hybrid functions are randomly separated into different subdivisions, which are replaced by using unimodal and multimodal functions, directly affecting the optimization performance. The results show that the optimization performance of GSA and PSO is not satisfactory for many hybrid functions, while ASO achieves a good optimization performance on them.

For composite functions f_{30}–f_{37}, Fig. 2.18 and Table 2.6 show that ASO performs a better convergence and obtains better statistical results than the others on functions f_{33}, f_{34}, f_{36}, and f_{37}. For functions f_{30} and f_{35}, ASO ranks only second to GSA and SA, respectively. In addition, ASO ranks third to WDO and GSA on function f_{31}. Even for function f_{32}, ASO is also superior to GA and PSO. Because the subfunctions from the basic functions contribute to too many local optima, it is very difficult for the algorithms to find the global optimum. It can clearly be shown that all the algorithms except for

ASO are trapped into local optima at different levels on functions f_{33}, f_{34}, f_{36}, and f_{37}. This is due to the fact that the repulsion of ASO contributes significantly to the exploration and avoids the local optima efficiently.

Tables 2.2–2.6 and Figs. 2.14–2.18 show that from the perspective of the convergence performance, ASO is highly competitive compared to its competitors on the benchmark functions, including unimodal, multimodal, low-dimensional, hybrid, and composition functions.

2.5 Conclusions

In this study, a novel physics-inspired metaheuristic algorithm, namely ASO algorithm, was developed for global optimization problems. ASO is inspired by the basic molecular dynamics to mathematically establish the atomic motion model, which is based on interaction and constraint forces. In ASO, each atom is affected by the attractive force or repulsive force from its neighbors and the constraint force from the atom with the best fitness value. The atomic motion follows Newton's second law. The attractive force encourages atoms to explore the entire search space extensively, and the repulsive force enables them to exploit the promising regions intensively (Zhao et al., 2019).

To test the effectiveness of ASO, five comparable algorithms were used on diverse benchmark sets including unimodal, multimodal, low-dimensional, hybrid, and composition functions at the same time for comparison. The results are verified regarding qualitative results, convergence preference, and overall preference. Based on the above results, analysis and discussion of the experiments and the application, the following conclusions can be drawn.

ASO performs a better self-adaptive convergence based on the different types of benchmark functions. The algorithm can successfully find the local regions around the promising solution.

ASO performs well for unimodal functions and shows a strong competitiveness for multimodal functions.

ASO can adeptly balance the explorative and exploitative search in dealing with hybrid and composition functions.

The simplicity of ASO and its few control parameters makes it easy to implement.

For future work, the binary version of ASO can be presented to deal with discrete problems, and its multiobjective version can also be proposed to solve multiobjective problems. ASO proposed here can be applied to other real-world problems.

References

Alder, B.J., Wainwright, T.E., 1959. Studies in molecular dynamics, I. general method. J. Chem. Phys. 31 (2), 459–466.
Ayman, A.A., 2011. Pid parameters optimization using genetic algorithm technique for electrohydraulic servo control system. Intell. Control. Autom. 2, 888–896.
Barker, G.F., 1870. A Text-Book of Elementary Chemistry: Theoretical and Inorganic. John P Morton & Co.
Bayraktar, Z., Komurcu, M., Bossard, J.A., Werner, D.H., et al., 2013. The wind driven optimization technique and its application in electromagnetics. IEEE Trans. Antennas Propag. 61 (5), 2745–2757.
Beyer, H.G., Schwefel, H.P., 2002. Evolution strategies-a comprehensive introduction. Nat. Comput. 1 (1), 3–52.
Birbil, Ş.İ., Fang, S.C., 2003. An electromagnetism-like mechanism for global optimization. J. Glob. Optim. 25 (3), 263–282.
Cuevas, E., González, M., 2013. An optimization algorithm for multimodal functions inspired by collective animal behavior. Soft. Comput. 17 (3), 489–502.
Digalakis, J., Margaritis, K., 2011. On benchmarking functions for genetic algorithms. Int. J. Comput. Math. 77, 481–506.
Dorigo, M., Maniezzo, V., Colorni, A., 1996. Ant system: optimization by a colony of cooperating agents. IEEE Trans. Syst. Man Cybern. B 26 (1), 29–41.
Erol, O.K., Eksin, I., 2006. A new optimization method: big bang–big crunch. Adv. Eng. Softw. 37 (2), 106–111.
Formato, R.A., 2007. Central force optimization. Prog. Electromagn. Res. 77, 425–491.
Goldstein, H., Poole, C.P., Safko, J.L., 2001. Classical Mechanics. Addison-Wesley, New Jersy.
Gray, C.G., Gubbins, K.E., 1984. Theory of Molecular Fluids, Volume 1: Fundamentals. Clarendon Press, Oxford.
Hasheminasab, F., Mehdipour, N., 2016. Molecular dynamics simulation of fluid sodium. Fluid Phase Equilib. 427, 161–165.
Holland, J.H., 1975. Adaptation in Natural and Artificial Systems. University of Michigan Press, Ann Arbor, MI.
Huan, T.T., Kulkarni, A.J., Kanesan, J., Huang, C.J., Abraham, A., 2017. Ideology algorithm: a socio-inspired optimization methodology. Neural Comput. & Applic. 28 (1), 845–876.
Juste, K.A., Kita, H., Tanaka, E., et al., 1999. An evolutionary programming solution to the unit commitment problem. IEEE Trans. Power Syst. 14 (4), 1452–1459.
Karaboga, D., Akay, B., 2007. A comparative study of artificial bee colony algorithm. Appl. Math. Comput. 214 (1), 108–132.
Kashan, A.H., 2009. League championship algorithm: a new algorithm for numerical function optimization. In: 2009 International Conference of Soft Computing and Pattern Recognition. IEEE, pp. 43–48.
Kenkel, J., Kelter, P.B., Hage, D.S., et al., 2000. Chemistry: An Industry-based Introduction with CD-ROM. CRC Press, Boca Raton, FL.
Kennedy, J., Eberhart, R., 1995. Particle swarm optimization. In: Proceedings of Sixth International Symposium on Micro Machine and Human Science (SMMHS-1995), pp. 39–43.
Kilymis, D.A., Delaye, J.M., Ispas, S., et al., 2016. Density effects on the structure of irradiated sodium borosilicate glass: a molecular dynamics study. J. Non-Cryst. Solids 432, 354–360.
Kirkpatrick, S., Gelatt, C.D., Vecchi, M.P., 1983. Optimization by simulated annealing. Science 220 (4598), 671–680.

Kourakos, G., Mantoglou, A., 2013. Development of a multi-objective optimization algorithm using surrogate models for coastal aquifer management. J. Hydrol. 479, 13–23.
Kumar, A., Joseph, S., Tsechansky, L., Privat, K., Schreiter, I.J., Schüth, C., Graber, E.R., 2018. Biochar aging in contaminated soil promotes Zn immobilization due to changes in biochar surface structural and chemical properties. Sci. Total Environ. 626, 953–961.
Kuo, H.C., Lin, C.H., 2013. Cultural evolution algorithm for global optimizations and its applications. J. Appl. Res. Technol. 11 (4), 510–522.
Lai, C., Shao, Q., Chen, X., Wang, Z., 2016. Flood risk zoning using a rule mining based on ant colony algorithm. J. Hydrol. 542, 268–280.
Lennard-Jones, J.E., 1924. On the determination of molecular fields. Proc. R. Soc. Lond. A 106 (738), 463–477.
Liang, J.J., Qu, B.Y., Suganthan, P.N., et al., 2014. Problem definitions and evaluation criteria for the CEC 2014 special session and competition on single objective real-parameter numerical optimization. In: Computational Intelligence Laboratory, Zhengzhou University, Zhengzhou China and Technical Report. Nanyang Technological University, Singapore. 2013.
Maitland, G.C., Rigby, M., Smith, E.B., Wakeham, W.A., et al., 1981. Intermolecular Forces: Their Origin and Determination. Clarendon Press, Oxford.
Mirjalili, S., Lewis, A., 2016. The whale optimization algorithm. Adv. Eng. Softw. 95, 51–67.
Moosavian, N., Roodsari, B.K., 2014. Soccer league competition algorithm: a novel meta-heuristic algorithm for optimal design of water distribution networks. Swarm Evol. Comput. 17, 14–24.
Moscato, P., Mendes, A., Berretta, R., 2007. Benchmarking a memetic algorithm for ordering microarray data. Biosystems 88 (1–2), 56–75.
Pál, K.F., 2006. Hysteretic optimization for the Sherrington–Kirkpatrick spin glass. Physica A 367, 261–268.
Pan, W.T., 2012. A new fruit fly optimization algorithm: taking the financial distress model as an example. Knowl.-Based Syst. 26, 69–74.
Passino, K.M., 2002. Biomimicry of bacterial foraging for distributed optimization and control. IEEE Control. Syst. 22 (3), 52–67.
Pishkenari, H.N., Mohagheghian, E., Rasouli, A., et al., 2016. Molecular dynamics study of the thermal expansion coefficient of silicon. Phys. Lett. A 380 (48), 4039–4043.
Poli, L., Rocca, P., Manica, L., Massa, A., 2010. Handling sideband radiations in time-modulated arrays through particle swarm optimization. IEEE Trans. Antennas Propag. 58 (4), 1408–1411.
Rahman, A., 1964. Correlations in the motion of atoms in liquid argon. Phys. Rev. 136 (2A), A405–A411.
Rao, R.V., Savsani, V.J., Vakharia, D.P., 2012. Teaching–learning-based optimization: an optimization method for continuous non-linear large scale problems. Inf. Sci. 183 (1), 1–15.
Rapaport, D.C., Blumberg, R.L., McKay, S.R., Christian, W., et al., 1996. The art of molecular dynamics simulation. Comput. Phys. 10 (5), 456.
Rashedi, E., Nezamabadi-Pour, H., Saryazdi, S., 2009. GSA: a gravitational search algorithm. Inf. Sci. 179 (13), 2232–2248.
Ray, T., Liew, K.M., 2003. Society and civilization: an optimization algorithm based on the simulation of social behavior. IEEE Trans. Evol. Comput. 7 (4), 386–396.
Rocca, P., Oliveri, G., Massa, A., 2011. Differential evolution as applied to electromagnetics. IEEE Antennas Propag. Mag. 53 (1), 38–49.
Ryckaert, J.P., Ciccotti, G., Berendsen, H.J.C., et al., 1977. Numerical integration of the cartesian equations of motion of a system with constraints: molecular dynamics of *n*-alkanes. J. Comput. Phys. 23 (3), 327–341.

Satapathy, S., Naik, A., 2016. Social group optimization (SGO): a new population evolutionary optimization technique. Complex Intell. Syst. 2 (3), 173–203.
Shahabinejad, H., Sohrabpour, M., 2017. A novel neutron energy spectrum unfolding code using particle swarm optimization. Radiat. Phys. Chem. 136, 9–16.
Simon, D., 2009. Biogeography-based optimization. IEEE Trans. Evol. Comput. 12 (6), 702–713.
Stone, A.J., 1996. The Theory of Intermolecular Forces. Clarendon Press, Oxford.
Thilak, K.D., Amuthan, A., 2018. Cellular automata-based improved ant colony-based optimization algorithm for mitigating ddos attacks in vanets. Futur. Gener. Comput. Syst. 82, 304–314.
Tuckerman, M.E., Berne, B.J., Martyna, G.J., et al., 1993. Efficient molecular dynamics and hybrid Monte Carlo algorithms for path integrals. J. Chem. Phys. 99 (4), 2796–2808.
Walker, S.M., King, A., 2005. What is Matter? Lerner Publications, Minneapolis, MN.
Wolpert, D.H., Macready, W.G., 1997. No free lunch theorems for optimization. IEEE Trans. Evol. Comput. 1 (1), 67–82.
Xu, Y., Cui, Z., Zeng, J., 2010. Social emotional optimization algorithm for nonlinear constrained optimization problems. In: International Conference on Swarm, Evolutionary, and Memetic Computing. Springer, Berlin, Heidelberg, pp. 583–590.
Yang, X.S., Deb, S., 2009. Cuckoo search via Lévy flights. In: Nature & Biologically Inspired Computing, NaBIC 2009, World Congress on IEEE, pp. 210–214.
Zhang, M., Luo, W., Wang, X., 2008. Differential evolution with dynamic stochastic selection for constrained optimization. Inf. Sci. 178 (15), 3043–3074.
Zhao, W., Wang, L., 2016. An effective bacterial foraging optimizer for global optimization. Inf. Sci. 329, 719–735.
Zhao, W., Wang, L., Zhang, Z., 2019. Atom search optimization and its application to solve a hydrogeologic parameter estimation problem. Knowl.-Based Syst. 163, 283–304.

CHAPTER 3

Engineering applications of atom search optimization algorithm

Contents

3.1 Introduction

In this part, the proposed atom search optimization (ASO) (Zhao et al., 2019) is successfully applied to parameter estimation for a chaotic system, circular antenna array design problem, and spread spectrum radar polyphase code design. The experiment results illustrate its superiority over its competitors.

3.2 Parameter estimation for chaotic system

3.2.1 Simulation model

A series of observed time data produced by a device is considered as the specific system, and a model used as the estimated system, in which the parameters need to be adjusted to make sure two systems are in synchronization. However, the time series of state variables of the estimated system are different from that of the specific system because of unknown system parameters. Our aim is to use ASO to search for a set of system parameters of the

New Optimization Algorithms and their Applications
https://doi.org/10.1016/B978-0-323-90941-9.00003-X

estimated system which make it synchronize with the specific system (Zhao and Wang, 2016).

The continuous nonlinear chaotic system is written by

$$\dot{\boldsymbol{x}} = f(\boldsymbol{x}, \boldsymbol{x}_0, \boldsymbol{\theta}_0) \tag{3.1}$$

where $\boldsymbol{x} = (x_1, x_2, \cdots, x_n) \in R^n$ indicates the state vector, $\boldsymbol{x}_0$ represents the initial state, and $\theta_0 = (\theta_{01}, \theta_{02}, \cdots, \theta_{0m})$ is the original parameters of the chaotic system; $\dot{\boldsymbol{x}}$ denotes the temporal derivative of $\boldsymbol{x}$. If the specific system form of Eq. (3.1) is known, then the estimated system can be built as

$$\dot{\boldsymbol{y}} = f(\boldsymbol{y}, \boldsymbol{x}_0, \boldsymbol{\theta}) \tag{3.2}$$

where $\boldsymbol{y} = (y_1, y_2, \cdots, y_n) \in R^n$ is the state vector and $\boldsymbol{\theta} = (\theta_1, \theta_2, \cdots, \theta_m)$ is the estimated parameters of the chaotic system, assuming that there exists a unidirectional coupling scheme that can keep the estimated system synchronized with the given chaotic system using any of its signal outputs, as long as the original parameters are equal to the estimated parameters. Thus, based on the measurable state vector $\boldsymbol{x} = (x_1, x_2, \cdots, x_n)$, the objective function is defined as

$$F = \sum_{i=0}^{T} [x_1(t) - y_1(t)]^2 + [x_2(t) - y_2(t)]^2 + \cdots + [x_n(t) - y_n(t)]^2 \tag{3.3}$$

where T is a series of discrete data. Therefore, the parameter identification of the estimated system is transformed into a minimization problem by using optimization algorithms to search for the suitable value of the parameter θ to minimize the objective function. Fig. 3.1 shows the block diagram of parameter identification for the chaotic system using optimization algorithms. The Lorenz system (Li and Yin, 2012; Zhao and Wang, 2016), a very notable chaotic system, is widely used in many studies and can be described as

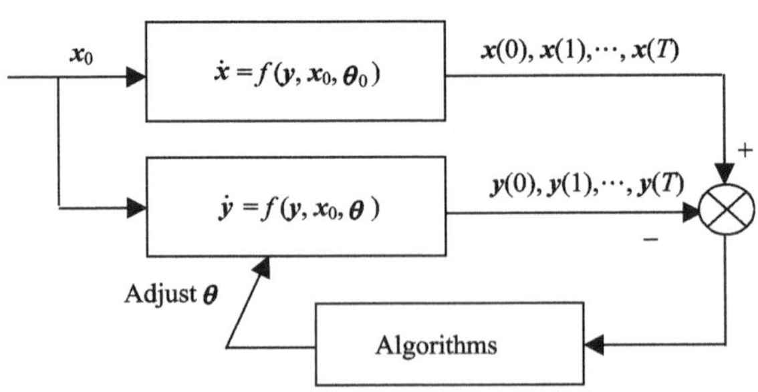

Fig. 3.1 Block diagram of parameter identification for chaotic system.

$$\begin{cases} \dot{x}_1 = \theta_{01}(x_1 - x_2) \\ \dot{x}_2 = \theta_{02}x_1 - x_1x_3 \\ \dot{x}_3 = x_1x_2 - \theta_{03}x_3 \end{cases} \quad (3.4)$$

where $\boldsymbol{x} = (x_1, x_2, x_3) \in R^3$ is the state vector and $\boldsymbol{\theta}_0 = (\theta_{01}, \theta_{02}, \theta_{03})$ are the original parameters. The Lorenz system is in the chaotic state when $\theta_{01} = 10$, $\theta_{02} = 28$, and $\theta_{03} = 8/3$. The dynamics trajectory of the Lorenz system is depicted in Fig. 3.2. The estimated Lorenz system is described as follows:

$$\begin{cases} \dot{y}_1 = \theta_1(y_1 - y_2) \\ \dot{y}_2 = \theta_2y_1 - y_1y_3 \\ \dot{y}_3 = y_1y_2 - \theta_3y_3 \end{cases} \quad (3.5)$$

where $\boldsymbol{y} = (y_1, y_2, y_3) \in R^3$ is the state vector and $\theta = (\theta_1, \theta_2, \theta_3)$ are the parameters to be estimated to attain synchronization with the Lorenz system with the original parameters.

3.2.2 Results and discussion

To implement the optimization for the estimated Lorenz system using ASO, the following control parameters are set: the depth weight is 50, the multiplier weight is 0.2, the maximum number of iterations is 200, and the population size is 20. For the simulation, the searching ranges for the estimated parameters are set at $9 \le \theta_1 \le 12$, $25 \le \theta_2 \le 30$, and $0 \le \theta_3 \le 3$. Moreover, the initial state vectors both $\boldsymbol{x}(0)$ and $\boldsymbol{y}(0)$ are all chosen as (0.1, 0.1, 0.1).

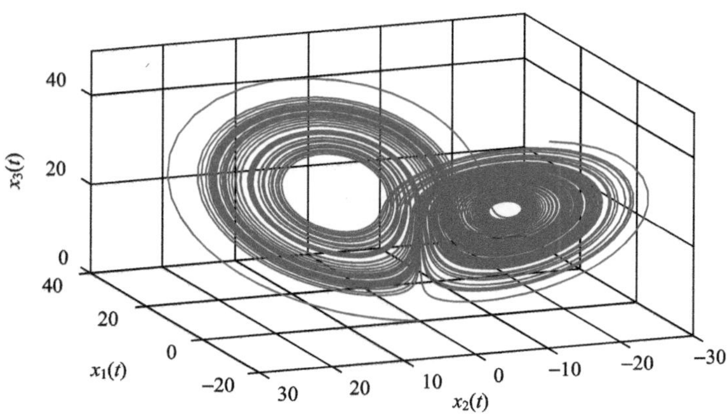

Fig. 3.2 Dynamics trajectory of Lorenz system.

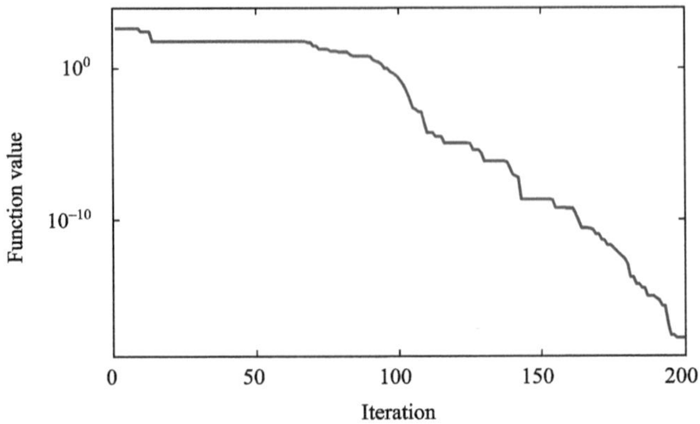

Fig. 3.3 Estimation process for the parameters of the Lorenz system.

A typical estimation process for the parameters of the Lorenz system is shown in Fig. 3.3. As shown in this figure, ASO perform a fast convergence rate for the global optimal solution. Fig. 3.4 shows a searching process for parameters θ_1, θ_2, and θ_3. Obviously, it can be seen from Fig. 3.4 that these three parameters all can converge to the true values rapidly. After 200 iterations, the final function value using ASO is $F = 1.579\text{E}-18$ with

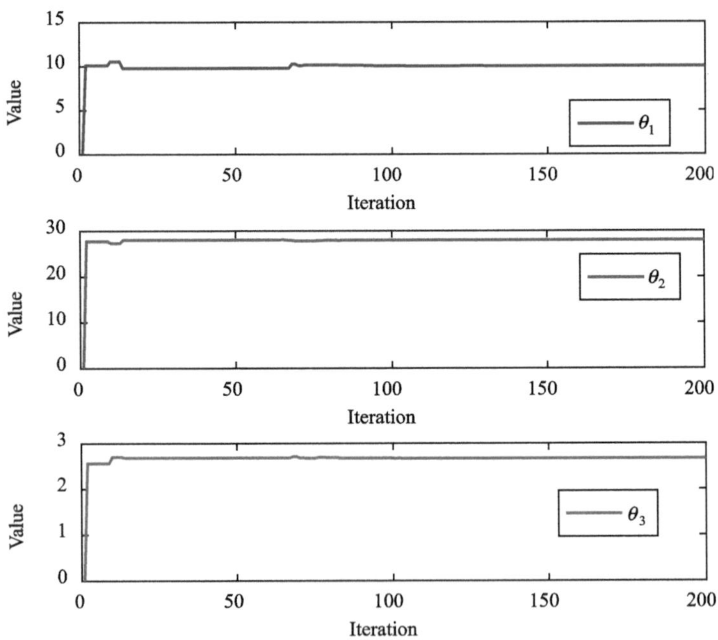

Fig. 3.4 Searching process for parameters θ_1, θ_2, and θ_3 using ASO.

$\theta_1 = 10.00000000003508$, $\theta_2 = 28.00000000000243$, and $\theta_3 = 2.666666666666645$. These results show ASO has a fast convergence speed and has good efficiency to achieve global optimization of parameters of a chaotic system.

3.3 Circular antenna array design problem

3.3.1 Problem description

Circular shaped antenna arrays are applied to various engineering fields (Dessouky et al., 2006a,b; Gurel and Ergul, 2008; Avishek et al., 2018; Pani et al., 2019). N antenna elements are spaced on a circle of radius r in the x–y plane and the antenna elements constitute a circular antenna array. Fig. 3.5 shows the geometry of circular antenna array. The array factor for the circular array is presented as follows (Das and Suganthan, 2012)

$$AF(\phi) = \sum_{n=1}^{N} I_n \exp\left\{jkr\left[\cos\left(\phi - \phi_{\text{ang}}^{n}\right) - \cos\left(\phi_0 - \phi_{\text{ang}}^{n}\right)\right] + \beta_n\right\} \quad (3.6)$$

where $\phi_{\text{ang}}^{n} = 2\pi(n-1)/N$is the angular position of the nth element on the x–y plane, $kr = nd$, k is the wave number, d is the angular spacing between elements, r is the radius of the circle defined by the antenna array, ϕ_0 is the direction of maximum radiation, ϕ is the angle of incidence of the plane wave, I_n is the current excitation, and β_n is the phase excitation of the nth element.

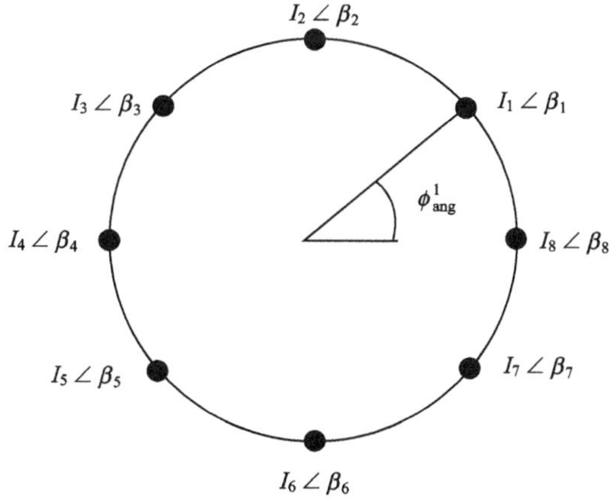

Fig. 3.5 Geometry of the circular antenna array (Das and Suganthan, 2012).

A symmetrical excitation of the circular antenna array is considered. The relations are given as below

$$\begin{aligned} I_{n/2+1}\angle\beta_{n/2+1} &= \text{conj}\ (I_1\angle\beta_1), \\ I_{n/2+2}\angle\beta_{n/2+2} &= \text{conj}\ (I_1\angle\beta_1), \\ &\cdots \\ I_n\angle\beta_n &= \text{conj}\ \left(I_{n/2}\angle\beta_{n/2}\right) \end{aligned} \tag{3.7}$$

So, the objective function is given as

$$\begin{aligned} OF = &|AR(\varphi_{\text{sll}}, \boldsymbol{I}, \boldsymbol{\beta}, \varphi_0)|/|AR(\varphi_{\max}, \mathbf{I}, \boldsymbol{\beta}, \varphi_0)| + 1/DIR(\varphi_0, \boldsymbol{I}, \boldsymbol{\beta}) \\ &+ |\varphi_0 - \varphi_{\text{des}}| + \sum_{k=1}^{N} |AR(\varphi_k, \boldsymbol{I}, \boldsymbol{\beta}, \varphi_0)| \end{aligned} \tag{3.8}$$

For this application, the instantiation is provided as below (Das and Suganthan, 2012):

(1) number of elements in circular array = 12;
(2) x_1 = Any string within bounds;
(3) null = [50, 120] in radians (no null control);
(4) phi_desired = 180°;
(5) distance = 0.5.

3.3.2 Results and discussion

The population size of 50 and maximum number of iterations of 1000 are considered. Fig. 3.6 shows the convergence curve for a various number of iterations. Fig. 3.7 shows the values of different elements in circular array for a various number of iterations. Based on these figures, it can be seen that ASO can provide a fast convergence rate and an optimal solution for this problem. Table 3.1 gives the optimal value of each element provided by ASO in a circular array.

3.4 Spread spectrum radar polyphase code design

3.4.1 Problem description

The spread spectrum radar polyphase code design problem is the minimization of the module of the biggest among the samples of the auto-correlation function which is related to the complex envelope of the compressed radar pulse at the optimal receiver output. This problem can be considered as a min-max nonlinear nonconvex optimization problem in continuous

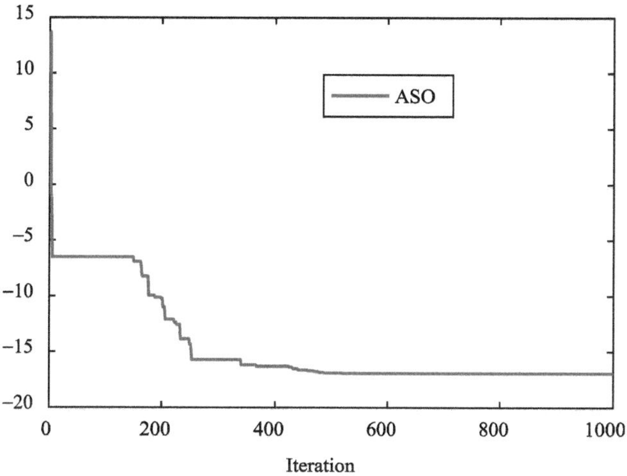

Fig. 3.6 Convergence curve for various number of iterations.

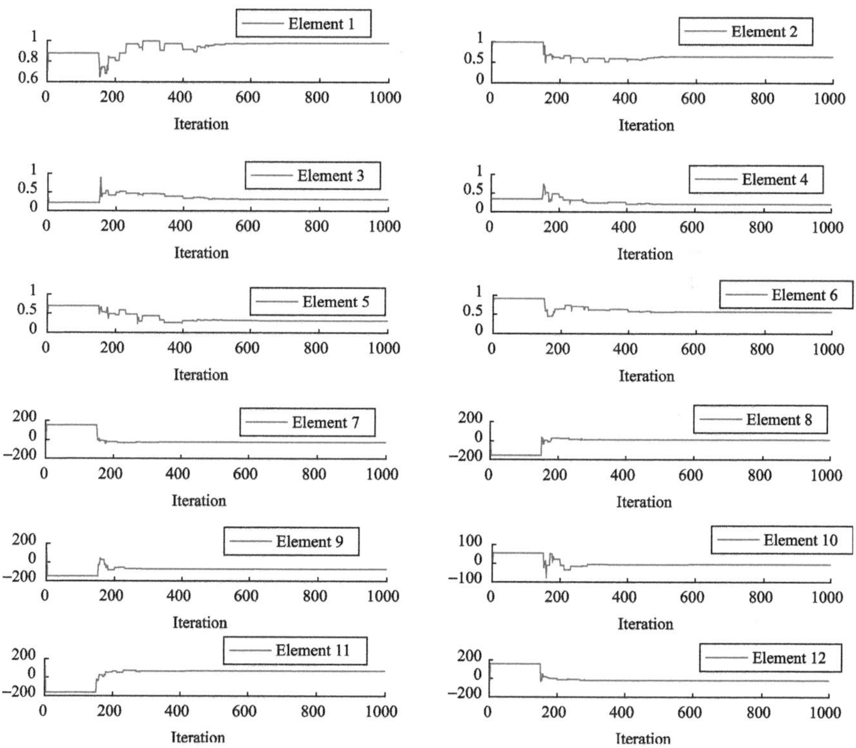

Fig. 3.7 Values of different elements in circular array for various number of iterations.

Table 3.1 Optimal value of each element provided by ASO in circular array.

Element	**Element 1**	**Element 2**	**Element 3**	**Element 4**	**Element 5**	**Element 6**
Value	0.975528	0.638072	0.309835	0.204040	0.312524	0.567210
Element	**Element 7**	**Element 8**	**Element 9**	**Element 10**	**Element 11**	**Element 12**
Value	−25.387095	12.227045	−72.129862	−5.776817	68.540695	−18.486118

variables and with numerous local optima. It can be expressed as follows (Das and Suganthan, 2012; Mladenović et al., 2003; Das et al., 2009):

$$\min_{x \in X} f(x) = \max\left[\phi_1(x), \cdots, \phi_{2m}(x)\right] \tag{3.9}$$

$$X = \left[(x_1, \cdots, x_n) \in R^n \middle| 0 \leq x_j \leq 2\pi, j = 1, \cdots, n\right] \tag{3.10}$$

where $m = 2n - 1$, and

$$\phi_{2i-1}(x) = \sum_{j=1}^{n} \cos\left(\sum_{k=|2i-j-1|+1}^{j} x_k\right), i = 1, \cdots, n \tag{3.11}$$

$$\phi_{2i}(x) = 0.5 + \sum_{j=i+1}^{n} \cos\left(\sum_{k=|2i-j|+1}^{j} x_k\right), i = 1, \cdots, n-1 \tag{3.12}$$

3.4.2 Results and discussion

To check the effectiveness of the ASO algorithm, some classic optimization algorithms were also used to solve this problem and their results are compared to the results by the ASO. The population size of 50 and maximum number of iterations of 1000 are considered for all the considered algorithms. Fig. 3.8 shows the convergence curve for various number of

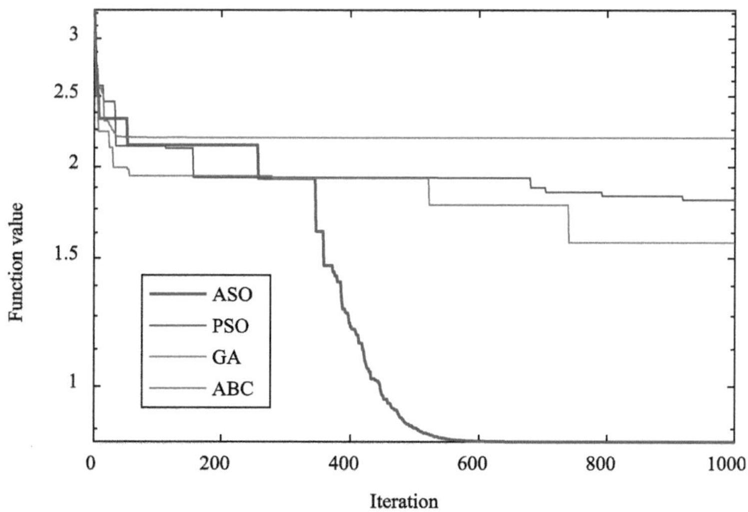

Fig. 3.8 The convergence curve for various number of iterations using different algorithms.

Table 3.2 Optimal solutions obtained using different algorithms.

Algorithms	ASO	PSO	GA	ABC
x_1	1.71641	1.02553	2.02654	3.84282
x_2	3.04256	5.95354	3.83231	5.63944
x_3	4.57876	3.46930	2.18407	3.71840
x_4	4.33689	5.79632	3.83941	2.84478
x_5	4.97041	4.52930	3.33459	4.82375
x_6	5.42520	4.33973	0.18372	4.20587
x_7	3.91513	1.41335	0.11911	6.02291
x_8	5.12911	3.48050	1.03232	2.11683
x_9	3.57836	1.99584	3.47345	0.97012
x_{10}	2.35872	4.73524	1.99180	4.90781
x_{11}	2.64514	4.93814	2.60498	2.31088
x_{12}	1.87116	0.94298	1.46409	2.60644
x_{13}	4.99852	5.48088	4.03092	3.60658
x_{14}	1.49900	4.71469	0.62201	5.25646
x_{15}	1.30922	6.11554	3.81852	2.55537
x_{16}	0.97581	0.82651	2.07021	0.91759
x_{17}	4.76633	0.88874	0.74399	0.59832
x_{18}	2.68486	4.45341	2.76354	5.02222
X_{19}	0.75114	4.21601	0.35444	4.72874
x_{20}	4.30844	2.82436	3.95479	2.04141
f_{min}	0.84025	1.80413	2.19480	1.57536

iterations using different algorithms. This figure shows that the ASO algorithm is better or competitive to the other optimization methods when solving this problem. Table 3.2 gives the optimal solutions and objective function values using PSO (Kennedy and Eberhart, 1995), GA (Holland, 1975), ABC (Akay and Karaboga, 2012) and ASO algorithms. It can be concluded that the ASO algorithm is more effective and potential in solving the spread spectrum radar polyphase code design problem.

3.5 Conclusions

To show effectiveness and practicability of ASO, several applications such as parameter estimation of a chaotic system, circular antenna array design, and spread spectrum radar polyphase code design problems were investigated in detail. The convincing results show ASO is superior to some classic and newly emerging algorithms in the literature when dealing with these complex problems. Therefore, ASO is a promising alternative to some classic optimizers extensively used in many real-world problems.

References

Akay, B., Karaboga, D., 2012. A modified artificial bee colony algorithm for real-parameter optimization. Inf. Sci. 192, 120–142.

Avishek, D., Mandal, D., Ghoshal, S.P., Kar, R., 2018. Concentric circular antenna array synthesis for side lobe suppression using moth flame optimization. Arch. Elektron. Ubertrag.: Electron. Commun. 86, 177–184.

Das, S., Suganthan, P.N., 2012. Problem Definitions and Evaluation Criteria for CEC 2011 Competition on Testing Evolutionary Algorithms on Real World Optimization Problems, Technical Report. pp. 1–42.

Das, S., Konar, A., Chakraborty, U.K., Abraham, A., 2009. Differential evolution with a neighborhood based mutation operator: a comparative study. IEEE Trans. Evol. Comput. 13 (3), 526–553.

Dessouky, M., Sharshar, H., Albagory, Y., 2006a. A novel tapered beamforming window for uniform concentric circular arrays. J. Electromagn. Waves Appl. 20 (14), 2077–2089.

Dessouky, M., Sharshar, A., Albagory, Y., 2006b. Efficient sidelobe reduction technique for small-sized concentric circular arrays. Prog. Electromagn. Res. 65, 187–200.

Gurel, L., Ergul, O., 2008. Design and simulation of circular arrays of trapezoidal-tooth log-periodic antennas via genetic optimization. Prog. Electromagn. Res. 85, 243–260.

Holland, J.H., 1975. Adaptation in Natural and Artificial Systems. University of Michigan Press, Ann Arbor, MI.

Kennedy, J., Eberhart, R., 1995. Particle swarm optimization. In: Proceedings of the 1995 IEEE International Conference on Neural Networks 194, pp. 2–8.

Li, X.T., Yin, M.H., 2012. Parameter estimation for chaotic systems using the cuckoo search algorithm with an orthogonal learning method. Chin. Phys. B 21 (5), 1–5.

Mladenović, N., Petrovic, J., Kovacevic-Vujicic, Cangalovic, M., 2003. Solving spreadspectrum radar polyphase code design problem by tabu search and variable neighborhood search. Eur. J. Oper. Res. 153, 389–399.

Pani, S., Tripathy, M.R., Kumar, A., 2019. Analysis of rectangular patch array antenna on circular plane for tera hertz applications. Optik 188, 302–307.

Zhao, W., Wang, L., 2016. An effective bacterial foraging optimizer for global optimization. Inf. Sci. 329, 719–735.

Zhao, W., Wang, L., Zhang, Z., 2019. Atom search optimization and its application to solve a hydrogeologic parameter estimation problem. Knowl.-Based Syst. 163, 283–304.

CHAPTER 4

Artificial ecosystem-based optimization algorithm

Contents

4.1 Introduction

Optimization for real-world problems is becoming an increasingly challenging field that has attracted considerable attention from researchers for several decades. A large number of scholars explore different methods to tackle all types of real-world optimization problems, and the methods most often used are numerical that usually adopt simple and ideal mathematical models. However, these methods generally need some gradient information to seek better solutions revolving around a special point in a local region. In addition, they are very sensitive to initial points, and especially when the considered problems have multiple or sharp peaks, the incorrect choice of initial points tends to make the search for the global optimum difficult and unstable (Adeli and Cheng, 1993). Recently numerous complex optimization problems have been increasingly emerging in different domains, where these problems often concern multiple decision variables, complex nonlinear constraints, and objective functions (Talbi, 2009; Wang et al., 2018). Therefore,

https://doi.org/10.1016/B978-0-323-90941-9.00004-1

these complex problems fail to be solved by traditional numerical methods with an acceptable time and solution precision. However, nature offers abundant sources for us including thoughts, inspiration, and concepts to develop artificial intelligent methods that are used to tackle these complex problems.

Over time, various nature-inspired optimization approaches have been proposed. These algorithms imitate behaviors of living organisms or natural phenomena for various problems (Mühlenbein et al., 1988; Geem et al., 2001; Hare et al., 2013; Krause et al., 2013). Genetic algorithm (GA), conceived by Holland (1975), evolves populations by imitating the survival of the fittest in nature; particle swarm optimization (PSO), developed by Kennedy and Eberhart (1995), models the movement of a population of individuals (e.g., birds); ant colony optimization (ACO), introduced by Dorigo and Maniezzo (Dorigo et al., 1996), is motivated from the foraging process of an ant colony; simulated annealing (SA), proposed by Kirkpatrick et al. (1983), is originated from the annealing process used in physical materials; and differential evolution (DE), presented by Storn and Price (1997), simulates the biological processes of genetic inheritance and survival of the fittest. These approaches are very popular in scientific fields of intelligent computing because they provide a better optimization performance especially for those nondifferentiable, noncontinuous, multimodal, and multidimensional problems. With their significant attention and applications, a large number of other optimization algorithms have been developed and applied to a range of fields with success (Milner et al., 2012; Ayman, 2011; Hamidreza and Karim, 2008; Zhang et al., 2016; Lai et al., 2016; Tarek et al., 2011; Nayak et al., 2018; Passino, 2002; Beyer and Schwefel, 2002; Vikhar, 2016; Corno et al., 1998; Eusuff and Lansey, 2003; Simon, 2009; Uymaz et al., 2015; Civicioglu, 2013; Pan, 2012; Akay and Karaboga, 2009; Oftadeh et al., 2010; Kiran, 2015; Mirjalili and Lewis, 2016; Saremi et al., 2017; Yang and Deb, 2009; Fathollahi-Fard et al., 2018; Kaveh and Farhoudi, 2013; Yang, 2010; Li et al., 2016; Zhao et al., 2019a,c; Mirjalili et al., 2017; Gandomi and Alavi, 2012; Cuevas et al., 2013; Askarzadeh, 2014; Rashedi et al., 2009; Formato, 2007; Bayraktar et al., 2013; Xie and Zeng, 2010; Abedinpourshotorban et al., 2016; Kaveh and Bakhshpoori, 2016; Shah-Hosseini, 2011; Chuang and Jiang, 2007; Shah-Hosseini, 2009; Kaveh and Dadras, 2017; Kaveh and Talatahari, 2010; Zheng et al., 2010; Javidy et al., 2015; Mirjalili and Hashim, 2012; Zheng, 2015; Flores et al., 2011; Tamura and Yasuda, 2011; Rao et al., 2012; Zarand et al., 2002; Shen and Li, 2009; Kripka and Kripka, 2008; Patel and Savsani, 2015; Eskandar et al., 2012; Moghaddam et al., 2012).

The successful application of these algorithms in solving some complex problems makes them valued by scientists and technicians.

4.2 Artificial ecosystem-based optimization

In this section, the inspiration of AEO is firstly introduced (Zhao et al., 2019b). Then its mathematical model is provided in detail.

4.2.1 Inspiration

The term "ecosystem", an intuitively appealing concept to most ecologists, was firstly introduced by Tansley in 1955. Owing to this appeal, the ecosystem concept has become increasingly important and popular (O'Neill et al., 1986). An ecosystem is a complex of living organisms, their physical environment, and all their interrelationships in a particular unit of space (Giannakos et al., 2016). An ecosystem can be classified into abiotic and biotic constituents. Abiotic constituents include sunlight, water, air, and other nonliving elements. Biotic constituents include all living organisms. The flow of energy and the cycling of nutrients are the chief driving forces maintaining the normal ecological order in an ecosystem, where all living organisms play a significant role. Living organisms can be categorized into producer, consumer, and decomposer.

Producers do not need to obtain energy from other organisms, and most producers are any type of green plant that obtains their food energy directly through the process of photosynthesis. During photosynthesis, carbon dioxide and water react together in the presence of sunlight to produce a sugar (glucose) and oxygen. Plants use this sugar to make many things, such as leaves, fruits, wood, and roots. Producers generally provide essential food for herbivore consumers and omnivore consumers. Consumers are animals that cannot make their food, thus they must feed on producers or other consumers to obtain energy and nutrients. Consumers can be categorized into carnivores, herbivores, and omnivores. Animals that eat only producers (plants) are called herbivores. Animals that eat both producers and other animals are called omnivores. Animals that eat only other animals are called carnivores. Decomposers are organisms that feed on both dead plants (producers) and animals (consumers) or on the waste from living organisms. Decomposers include most bacteria and fungi. When an organism dies, decomposers break down the remains and convert them into simple molecules such as carbon dioxide, water, and minerals. Then these energy forms may be absorbed by producers to again make sugar and oxygen through photosynthesis, starting the cycle all over again.

Producers, consumers, and decomposers interacting with one another compose a food chain. A food chain describes who feeds whom in an ecosystem, where every life form needs to obtain food energy to achieve the movement of nutrients. A food chain is a possible connected path for energy and nutrients to flow through an ecosystem, and it shows the different levels of eating within an ecosystem. Many overlapped food chains form a food web that depicts how the food chains are interconnected. A food web is very complex because most consumers feed on producers and/or different types of consumers and also probably become food for other different types of consumers. Producers are generally at the beginning of feed webs and most food chains. Consumers are the most complex of the three types of organisms. Fig. 4.1 depicts the flow of energy in an ecosystem. The figure shows that the energy always flows from the living organisms with high energy towards the living organisms with low energy. Typically, the ecosystem develops this mechanism of energy transfer as a strategy to maintain the stability of species, and it is able to keep the ecological balance over the long term. Therefore, there are reasons to believe that this mechanism of energy transfer can shape and sustain a sound and stable ecosystem.

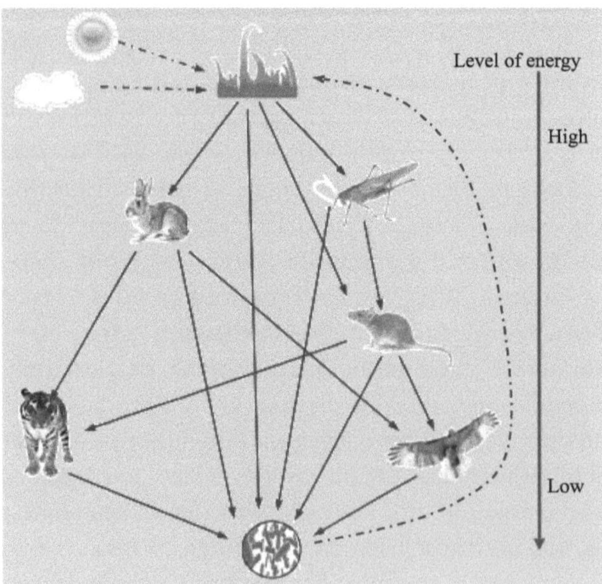

Fig. 4.1 Flow of energy in an ecosystem.

4.2.2 Artificial ecosystem-based optimization

Based on the previous discussion, our artificial ecosystem-based optimization (AEO) algorithm employs three operators: production, consumption, and decomposition. The first operator is mainly to enhance the balance between exploration and exploitation. The second operator is used to improve the exploration of the algorithm. For the third operator, it is proposed to promote the exploitation of the algorithm. AEO generally follows several rules to search for a solution:

(1) The ecosystem as a population includes three types of organisms: producer, consumer, and decomposer.
(2) There is only one producer as an individual in a population.
(3) There is only one decomposer as an individual in a population.
(4) The other individuals of a population are consumers, each of which is chosen as a carnivore, a herbivore, or an omnivore with the same probability.
(5) The energy level of each individual in a population is evaluated by its function fitness value. The population is sorted in order of descending function fitness value, so a higher function fitness value indicates the higher energy level for a minimization problem.

Fig. 4.2 shows an ecosystem in AEO. In this ecosystem, all individuals are sorted in order of descending function fitness value, and the black arrows represent the flow direction of energy. The worst individual x_1 (the highest function fitness value) and the best individual x_n (the lowest function fitness value) are a producer and a decomposer, respectively. The others are consumers, in which we suppose x_2 and x_5 are herbivores, x_3 and x_7 are omnivores, and x_4 and x_6 are carnivores.

4.2.2.1 Production

In an ecosystem, the producer may generate food energy with carbon dioxide, water, and sunlight, as well as the nutrition provided by the decomposer. Likewise, in AEO, the producer (the worst individual) in a population needs to be updated by the low and upper limits of search space and the decomposer (the best individual), and this updated individual will guide other individuals, including the herbivore and omnivore, in the population to search for different regions. The production operator allows AEO to randomly produce a new individual replacing the previous one between the best individual (x_n) and an individual randomly generated in the search space (x_{rand}). The mathematical model of the production operator is represented as follows:

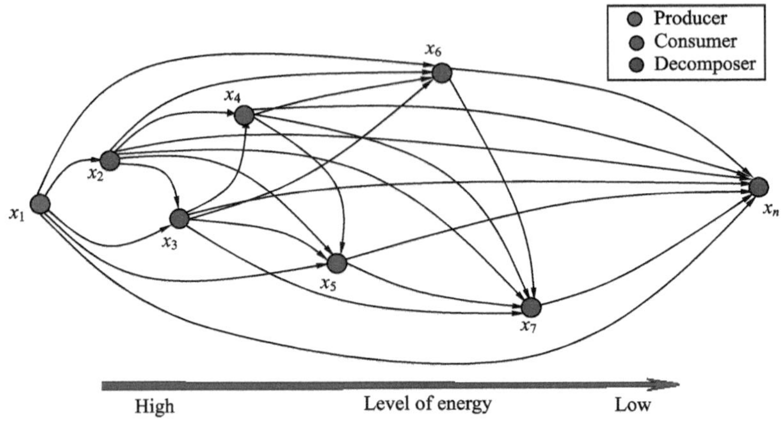

Fig. 4.2 An ecosystem in an artificial ecosystem-based optimization algorithm.

$$x_1(t+1) = (1-a)x_n(t) + ax_{\text{rand}}(t) \tag{4.1}$$

$$a = (1 - t/T)r_1 \tag{4.2}$$

$$x_{\text{rand}} = \mathbf{r}(U - L) + L \tag{4.3}$$

where n is the size of a population, T is the maximal number of iterations, L and U are the lower and upper limits, respectively, r_1 is a random number within the range of [0,1], $\mathbf{r}$ is a random vector within the range of [0, 1], a is a linear weight coefficient, and x_{rand} is the position of an individual randomly produced in the search space. In Eq. (4.1), the weight coefficient a is used to linearly drift the individual from a position randomly produced towards the position of the best individual as the iterations increase. As Eq. (4.1) shows, in the early iterations, $x_1(t+1)$ may guide the other individuals to perform an exploration in the search space extensively; in the later iterations, $x_1(t+1)$ may guide the other individuals to perform the exploitation in a region around x_n intensively.

This production operator in two-dimensional and three-dimensional spaces using Eq. (4.1) can be simulated, which is illustrated in Fig. 4.3, and this production behavior performs 20 iterations. As shown in the figure, this operator offered by Eq. (4.1) gradually moves an individual randomly generated towards the best individual. After 20 iterations, the individual reaches the position of the best individual.

4.2.2.2 Consumption

After the producer accomplishes its production operator, all the consumers may perform the consumption operator. To obtain food energy, each

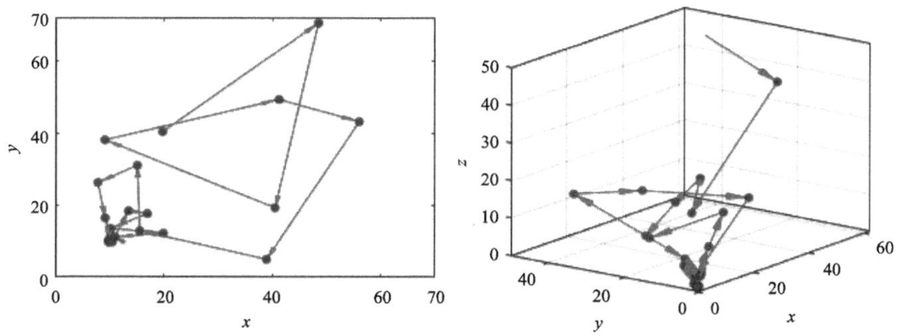

Fig. 4.3 Behaviors of production in two-dimensional and three-dimensional spaces during 20 iterations.

consumer may eat either a randomly-chosen consumer with the lower level of energy, or a producer, or both. Levy flight, as a mathematic operator, mimics food searching of many animals such as the cuckoo, bumblebee, deer, and lion. Levy flight is a random walk which can efficiently explore the search space as the length of some steps is much longer in the long run, showing that it is promising in finding the global optimum. Therefore, Levy flight is often added to nature-inspired algorithms to enhance their optimization performance (Yang and Deb, 2013; Viswanathan et al., 1996; Brown et al., 2007). However, there are two drawbacks to this behavior, including complexity and several parameters to be tuned. At a point, a simple, parameter-free random walk with the feature of Levy flight, called a consumption factor, is proposed. It is defined as follows:

$$C = \frac{1}{2}\frac{v_1}{|v_2|} \tag{4.4}$$

$$v_1 \sim N(0,1),\ v_2 \sim N(0,1) \tag{4.5}$$

where $N(0, 1)$ is a normal distribution with the mean (=0) and the standard (=1) deviation. The sample trajectories of the consumption factor in two-dimensional and three-dimensional spaces during 2000 iterations are shown in Fig. 4.4. It can be found that these random walks generally become crowded around a central point and occasionally permit a long step far from the previous point. It allows AEO to have the opportunity to avoid local extrema and to explore the entire search space. Therefore, this consumption factor may assist each consumer to hunt for food, while different types of consumers adopt different consumption strategies.

Herbivore: if a consumer is randomly chosen as an herbivore, it only eats the producer. As illustrated in Fig. 4.2, both consumers x_2 and x_5 are

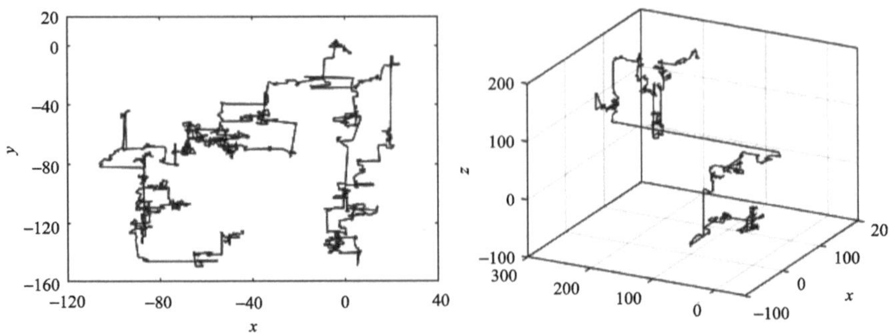

Fig. 4.4 Sample trajectories of consumption factor in two-dimensional and three-dimensional spaces during 2000 iterations.

herbivores which only consume the producer x_1. To mathematically model this consumption behavior of herbivores, the equation is presented as:

$$x_i(t+1) = x_i(t) + C \cdot [x_i(t) - x_1(t)], \ i \in [2, \cdots, n], \tag{4.6}$$

Carnivore: if a consumer is randomly chosen as a carnivore, it can only eat randomly a consumer with the higher energy level. In Fig. 4.2, the consumer x_6 is a carnivore, so it needs to randomly choose a consumer for food from the individuals x_2 to x_5 that all have higher energy levels than the consumer x_6. The equation modeling the consuming behavior of a carnivore is:

$$\begin{cases} x_i(t+1) = x_i(t) + C \cdot \left[x_i(t) - x_j(t)\right], \ i \in [3, \cdots, n] \\ j = \text{rand}i[(2, i-1)] \end{cases} \tag{4.7}$$

Omnivore: if a consumer is randomly chosen as an omnivore, it can eat both a consumer with the higher energy level randomly and a producer. In Fig. 4.2, the consumer x_7 is an omnivore, so it needs to eat both the producer x_1 and a consumer randomly chosen from the individuals x_2 to x_6 that have the higher energy levels than x_7. The mathematical equation modeling the consuming behavior of an omnivore can be expressed as:

$$\begin{cases} x_i(t+1) = x_i(t) + C \cdot \left\{r_2 \cdot [x_i(t) - x_1(t)] + (1 - r_2)\left[x_i(t) - x_j(t)\right]\right\}, \ i \in [3, \cdots, n] \\ j = \text{rand}i[(2, i-1)] \end{cases} \tag{4.8}$$

where r_2 is a random number within the range of [0, 1].

For this consumption operator, AEO updates the position of a search individual with respect to either the worst individual or a randomly-chosen individual in a population, or both. This behavior tends to emphasize the exploration and allows AEO to perform a global search.

4.2.2.3 Decomposition

Decomposition is a very vital process in terms of the functioning of an ecosystem, and it provides essential nutrients for the growth of the producer. During the decomposition, when each individual in the population dies, the decomposer will decay or chemically break down the remains. In order to mathematically model this behavior, the decomposition factor D and the weight coefficients e and h are designed. Hence, the position of the ith individual x_i in a population can be updated by the position of the decomposer x_n via these parameters D, e, and h. It allows the next position of each individual to spread around the best individual (the decomposer), showing the exploitation to some extent. The equation expressing this decomposition behavior is as follows:

$$x_i(t+1) = x_n(t) + D \cdot [e \cdot x_n(t) - h \cdot x_i(t)], \ i \in [1, \cdots, n] \quad (4.9)$$

$$D = 3u, \ u \sim N(0, 1) \quad (4.10)$$

$$e = r_3 \cdot \text{rand}i([1, 2]) - 1 \quad (4.11)$$

$$h = 2 \cdot r_3 - 1 \quad (4.12)$$

Fig. 4.5 shows that an individual samples 200 times in two-dimensional and three-dimensional spaces respectively according to Eq. (4.9). As shown in Fig. 4.5, most sampled points randomly distribute in the distance between the current individual x_i and the individual x_n, and the distribution of the sampled points become sparse as the distance between these two individuals reduces. Meanwhile, it can be seen that from this figure, a few of sampled points tend to randomly sample some regions far away from the individual x_n. This observation is able to support exploitation and avoidance of the local extrema.

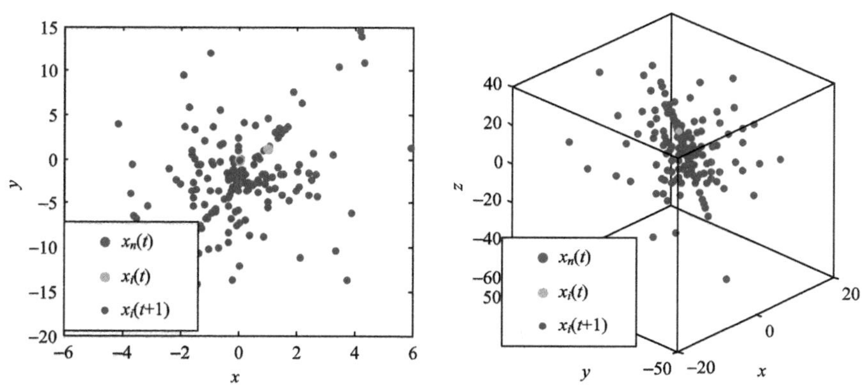

Fig. 4.5 Behaviors of consumption in two-dimensional and three-dimensional spaces during 200 samples.

AEO starts the optimization by generating a population randomly. At each iteration, the first search individual updates its position based on Eq. (4.1), and for the other individuals, there is the same probability to choose among Eqs. (4.6)–(4.8) to update their positions. The exception is if an individual is given a better function value. Then each individual updates its position based on Eq. (4.9). If an individual goes outside of the lower or upper boundaries during the updating process, it will be randomly generated in the search space. All the updates are interactively preformed until the AEO algorithm is satisfied with a termination criterion. Finally, the solution of the best individual found so far is returned. The pseudo code of AEO algorithm is described in Fig. 4.6.

4.3 Results and discussion

```
Randomly initialize an ecosystem X_i (solutions) and caculate the fitness Fit_i,
and X_best= the best solution found so far.
While the stop criterion is not satisfied do
        //Production//
    For individual X_1, update its solution using Eq. (4.1).
        //Consumption//
    For individual X_i (i=2,...,n),
        // Herbivore //
        If rand<1/3 then update its solution using Eq. (4.6),
        // Omnivore //
        Else If 1/3≤rand≤2/3 then update its solution using Eq. (4.7),
        // Carnivore //
            Else update its solution using Eq. (4.8),
          End If.
        End If.
     Caculate the fitness of each individual.
    Update the best solution found so far X_best.
        // Decomposition//
    Update the position of each individual using Eq. (4.9).
    Caculate the fitness of each individual.
    Update the best solution found so far X_best.
End While.
Return X_best.
```

Fig. 4.6 Pseudo code of AEO algorithm.

The numerical efficiency of AEO is analyzed on 31 popular benchmark functions to extensively evaluate its performance (Zhao et al., 2019b). These functions are described as unimodal functions, multimodal functions, low-dimensional functions, and composite function, which are listed in Appendix Tables A.1–A.3 and A.5, respectively. The unimodal functions are used to verify the exploitation of algorithms. The multimodal and low-dimensional functions are used to check the exploration of algorithms. The composite functions are constructed by the basic and hybrid functions, so they tend to check local optima avoidance of algorithms. The details of these composite functions from the CEC 2014 special session are available in Liang et al. (2013). Meanwhile, all these benchmark functions can be used to analyze the convergence behaviors of algorithms.

For these benchmark functions, the performance of AEO is compared with those of six stochastic optimizers, namely GA (Holland, 1975), PSO (Kennedy and Eberhart, 1995), DE (Storn and Price, 1997), CS (Yang and Deb, 2009), ABC (Akay and Karaboga, 2009), and GSA (Rashedi et al., 2009). Although a number of variants based on these or entirely new algorithms have been developed, they are not well explained in literature and are documented with incomplete parameter settings, rendering the exact replication of experiments and results nearly infeasible (Jain et al., 2019). Therefore, the standard versions of optimization algorithms are used for comparison. As we know, it is difficult for each function to search for a set of appropriate parameters to fit each algorithm. Therefore, for simplicity, for each algorithm, a set of fixed parameters are used to analyze its optimization performance on all benchmark functions. The initial parameters of different algorithms are provided as follows.

(1) PSO: inertia weight linearly reduces from 0.9 to 0.2, acceleration coefficients $c_1 = 2$ and $c_2 = 2$;

(2) GA: decreasing coefficient $\gamma = 20$, mutation rate $p_m = 0.2$, crossover rate $p_c = 0.8$ and crossover adopts roulette wheel method;

(3) DE: mutation factor $F = 0.5$ and crossover rate $C = 0.5$;

(4) CS: mutation probability $p_a = 0.25$;

(5) GSA: initial gravitational constant $G_0 = 100$ and decreasing coefficient $a = 20$.

4.3.1 Analysis of exploitation capability

Each of unimodal functions only has one global optimum and has no local optima, so they are generally used to test the exploitation of algorithms.

Table 4.1 shows the comparisons of optimization results offered by different optimizers on unimodal functions. "Mean" indicates the mean of best-so-far solution, "Std" indicates the standard deviation of best-so-far solution, and "Best" indicates the minimum of best-so-far solution. According to Table 4.1, the results of AEO are superior to those of the other metaheuristic algorithms. AEO has obtained the best results in terms of the mean, standard deviation, and minimum of best-so-far solution for all the unimodal functions. In particular, for the exploitation, CS provides the worst results on functions f_1 and f_2, ABC provides the worst results on functions f_3, f_4, and f_5, and, PSO and DE perform the worst for functions f_6 and f_7, respectively. It is clear that AEO shows the best exploitation capability of all the investigated optimizers. This merit results from the exploitation behavior in AEO previously discussed.

4.3.2 Analysis of exploration capability

Differing from unimodal functions, multimodal functions have an exponentially increasing number of local optima. Therefore, these functions are more suitable to test the exploration capability of algorithms. The comparisons of optimization results offered by different optimizers for multimodal functions are listed in Table 4.2, in which the best results are marked in bold. As this table shows, AEO performs the best of all the algorithms for all multimodal functions except function f_{13}. However, GSA provides the worst results on functions f_8 and f_{11}. ABC performs the worst on functions f_{12} and f_{13}. In addition, DE and CS also perform the worst for functions f_8 and f_9, respectively. Table 4.3 shows the comparisons of optimization results offered by different optimizers on low-dimensional multimodal functions. The results of AEO are not inferior to those of the other algorithms except for functions f_{15} and f_{20}. Although AEO fails to achieve the best results for these two functions, it only ranks behind CS and DE for function f_{15}, and performs better than PSO for function f_{20}. These results show that AEO provides a very good exploration capability. This is because the exploration operator is integrated into AEO.

4.3.3 Analysis of avoidance of local optima

Functions f_{24}–f_{31} are composite functions, their different variables are randomly separated into various subdivisions constructed by the basic and hybrid functions. This allows these functions to generate numerous local optima. Also, the global optimum is able to randomly shift from one position

Table 4.1 Comparisons of results for unimodal functions.

Function	Index	AEO	PSO	GA	DE	CS	GSA	ABC
f_1	Mean	**0.000E+00**	2.148E−04	9.247E−03	3.636E−14	9.675E−03	2.195E−17	2.363E−03
	Std	0.000E+00	2.247E−04	3.778E−03	6.064E−14	4.518E−03	6.379E−18	1.525E−03
	Best	0.000E+00	2.985E−06	3.206E−03	1.264E−15	3.270E−03	1.146E−17	7.092E−04
f_2	Mean	**5.248E−289**	2.957E−04	2.126E−02	4.385E−08	1.404E+00	2.283E−08	2.317E−04
	Std	0.000E+00	2.307E−04	5.609E−03	2.530E−08	5.609E−01	3.490E−09	1.530E−04
	Best	3.509E−301	4.155E−05	1.264E−02	1.494E−08	6.309E−01	1.647E−08	8.106E−05
f_3	Mean	**0.000E+00**	2.844E+03	1.053E+03	5.694E+00	4.725E+02	2.216E+02	9.562E+03
	Std	0.000E+00	1.343E+03	3.489E+02	3.910E+00	1.096E+02	7.067E+01	1.750E+03
	Best	0.000E+00	1.143E+03	3.967E+02	9.258E−01	2.900E+02	8.526E+01	5.926E+03
f_4	Mean	**2.958E−291**	1.736E+01	1.046E+00	9.172E+00	3.248E+00	3.451E−09	2.453E+01
	Std	0.000E+00	3.624E+00	2.897E−01	3.999E+00	8.547E−01	7.445E−10	2.284E+00
	Best	1.433E−304	1.089E+01	5.211E−01	2.184E+00	1.597E+00	2.258E−09	1.981E+01
f_5	Mean	**1.920E+01**	9.473E+01	9.856E+01	3.000E+01	3.861E+01	2.669E+01	5.470E+02
	Std	5.460E−01	7.896E+01	5.845E+01	1.765E+01	1.029E+01	2.671E+00	2.099E+02
	Best	1.743E+01	7.618E+00	1.299E+01	4.013E+00	2.872E+01	2.571E+01	2.397E+02
f_6	Mean	**0.000E+00**	1.333E−01	**0.000E+00**	1.333E−01	**0.000E+00**	**0.000E+00**	**0.000E+00**
	Std	0.000E+00	4.342E−01	0.000E+00	4.342E−01	0.000E+00	0.000E+00	0.000E+00
	Best	0.000E+00	0.000E+00	0.000E+00	0.000E+00	0.000E+00	0.000E+00	0.000E+00
f_7	Mean	**6.319E−05**	5.644E−02	4.473E−02	2.146E−01	3.085E−02	1.913E−02	9.528E−02
	Std	6.815E−05	2.032E−02	1.355E−02	7.238E−02	7.932E−03	6.870E−03	2.385E−02
	Best	3.848E−06	1.915E−02	1.510E−02	1.158E−01	1.381E−02	7.731E−03	5.285E−02

Table 4.2 Comparisons of results for multimodal functions.

Function	Index	AEO	PSO	GA	DE	CS	GSA	ABC
f_8	Mean	**−1.146E+04**	−5.139E+03	−6.820E+03	−5.310E+03	−8.692E+03	−2.639E+03	−5.124E+03
	Std	3.089E+02	5.777E+02	5.769E+02	6.617E+02	2.352E+02	4.351E+02	4.607E+02
	Best	−1.200E+04	−6.951E+03	−7.832E+03	−7.387E+03	−9.032E+03	−4.189E+03	−6.527E+03
f_9	Mean	**0.000E+00**	3.088E+01	1.240E+01	1.650E+02	8.323E+01	1.585E+01	1.578E+02
	Std	0.000E+00	8.884E+00	2.875E+00	1.751E+01	1.310E+01	3.848E+00	2.130E+01
	Best	0.000E+00	1.593E+01	7.967E+00	1.282E+02	5.889E+01	7.960E+00	1.111E+02
f_{10}	Mean	**8.882E−16**	7.599E−03	2.018E−02	5.413E−08	4.152E+00	3.443E−09	5.329E−02
	Std	0.000E+00	8.229E−03	4.771E−03	2.618E−08	1.487E+00	5.587E−10	4.047E−02
	Best	8.882E−16	9.453E−04	1.408E−02	1.812E−08	2.016E+00	2.655E−09	1.242E−02
f_{11}	Mean	**0.000E+00**	1.477E−02	2.158E−02	2.054E−03	9.626E−02	4.264E+00	1.631E−01
	Std	0.000E+00	1.361E−02	1.021E−02	3.888E−03	4.183E−02	1.588E+00	1.152E−01
	Best	0.000E+00	1.010E−04	6.839E−03	7.994E−15	2.609E−02	1.893E+00	8.277E−03
f_{12}	Mean	**2.016E−18**	3.758E−01	2.977E−05	6.911E−03	1.101E+00	3.399E−02	1.501E+01
	Std	4.097E−18	7.257E−01	3.014E−05	2.630E−02	3.172E−01	5.361E−02	4.671E+00
	Best	4.302E−20	1.435E−04	7.861E−06	1.335E−16	5.652E−01	5.162E−20	7.760E+00
f_{13}	Mean	1.831E−03	1.909E−01	6.310E−04	5.306E−02	1.343E−01	**2.041E−18**	3.626E+01
	Std	4.165E−03	3.879E−01	4.234E−04	2.886E−01	6.310E−02	5.131E−19	1.819E+01
	Best	1.800E−17	2.751E−03	3.591E−05	3.896E−15	6.292E−02	1.328E−18	1.294E+01

Table 4.3 Comparisons of results for low-dimensional functions.

Function	Index	AEO	PSO	GA	DE	CS	GSA	ABC
f_{14}	Mean	**9.980E−01**	**9.980E−01**	3.855E+00	**9.980E−01**	**9.980E−01**	3.473E+00	**9.980E−01**
	Std	0.000E+00	0.000E+00	2.766E+00	0.000E+00	0.000E+00	2.529E+00	6.916E−06
	Best	9.980E−01	9.980E−01	9.980E−01	9.980E−01	9.980E−01	9.980E−01	9.980E−01
f_{15}	Mean	3.543E−04	3.666E−04	1.175E−03	**3.075E−04**	**3.075E−04**	2.358E−03	5.097E−04
	Std	2.351E−04	1.908E−04	1.501E−03	1.420E−19	3.917E−08	1.143E−03	4.988E−05
	Best	3.075E−04	3.075E−04	3.207E−04	3.075E−04	3.075E−04	9.913E−04	4.036E−04
f_{16}	Mean	**−1.032E+00**	**−1.032E+00**	**−1.032E+00**	**−1.032E+00**	**−1.032E+00**	**−1.032E+00**	**−1.032E+00**
	Std	6.775E−16	6.775E−16	4.277E−10	6.775E−16	6.775E−16	5.532E−16	6.649E−16
	Best	−1.032E+00	−1.032E+00	−1.032E+00	−1.032E+00	−1.032E+00	−1.032E+00	−1.032E+00
f_{17}	Mean	**3.979E−01**	**3.979E−01**	**3.979E−01**	**3.979E−01**	**3.979E−01**	**3.979E−01**	**3.979E−01**
	Std	0.000E+00	0.000E+00	1.225E−08	0.000E+00	0.000E+00	0.000E+00	6.751E−10
	Best	3.979E−01	3.979E−01	3.979E−01	3.979E−01	3.979E−01	3.979E−01	3.979E−01
f_{18}	Mean	**3.000E+00**	**3.000E+00**	**3.000E+00**	**3.000E+00**	**3.000E+00**	**3.000E+00**	**3.000E+00**
	Std	2.070E−15	2.188E−15	1.837E−07	2.030E−15	1.882E−15	1.690E−15	2.861E−11
	Best	3.000E+00	3.000E+00	3.000E+00	3.000E+00	3.000E+00	3.000E+00	3.000E+00
f_{19}	Mean	**−3.863E+00**	**−3.863E+00**	**−3.863E+00**	**−3.863E+00**	**−3.863E+00**	**−3.863E+00**	**−3.863E+00**
	Std	2.710E−15	2.710E−15	1.454E−09	2.710E−15	2.710E−15	2.449E−15	1.668E−15
	Best	−3.863E+00	−3.863E+00	−3.863E+00	−3.863E+00	−3.863E+00	−3.863E+00	−3.863E+00

Continued

Table 4.3 Comparisons of results for low-dimensional functions—cont'd

Function	Index	AEO	PSO	GA	DE	CS	GSA	ABC
f_{20}	Mean	−3.259E+00	−3.254E+00	−3.298E+00	−3.286E+00	**−3.322E+00**	**−3.322E+00**	**−3.322E+00**
	Std	6.033E−02	6.042E−02	4.837E−02	5.542E−02	1.917E−13	1.355E−15	1.525E−15
	Best	−3.322E+00	−3.322E+00	−3.322E+00	−3.322E+00	−3.322E+00	−3.322E+00	−3.322E+00
f_{21}	Mean	**−1.015E+01**	−6.131E+00	−6.993E+00	−9.985E+00	**−1.015E+01**	−6.976E+00	−1.011E+01
	Std	7.227E−15	2.836E+00	3.700E+00	9.224E−01	3.803E−14	3.549E+00	1.683E−01
	Best	−1.015E+01	−1.015E+01	−1.015E+01	−1.015E+01	−1.015E+01	−1.015E+01	−1.015E+01
f_{22}	Mean	**−1.040E+01**	−8.218E+00	−9.225E+00	**−1.040E+01**	**−1.040E+01**	**−1.040E+01**	**−1.040E+01**
	Std	1.512E−15	2.979E+00	2.686E+00	1.745E−15	2.795E−14	0.000E+00	3.529E−14
	Best	−1.040E+01	−1.040E+01	−1.040E+01	−1.040E+01	−1.040E+01	−1.040E+01	−1.040E+01
f_{23}	Mean	**−1.054E+01**	−7.717E+00	−9.340E+00	**−1.054E+01**	**−1.054E+01**	**−1.054E+01**	**−1.054E+01**
	Std	1.745E−15	3.251E+00	2.731E+00	1.807E−15	4.200E−12	1.682E−15	8.095E−14
	Best	−1.054E+01	−1.054E+01	−1.054E+01	−1.054E+01	−1.054E+01	−1.054E+01	−1.054E+01

to another position during the iterations. Therefore, composite functions are very suitable for evaluating local optima avoidance which is able to balance the exploration and exploitation. The comparisons of optimization results offered by different optimizers on composite functions are provided in Table 4.4. As this table shows, the results of AEO are superior to all other algorithms for half of the composite functions. This reveals that AEO has high local optima avoidance resulting from an appropriate balance between exploration and exploitation. Such capability originates from the fact the production operator guides individuals to search from a randomly generated region far away from the best individual to a region close to the best individual as the iterations increase.

4.3.4 Analysis of convergence behavior

In this section, two functions with two dimensions are chosen to investigate the convergence behavior of AEO. The first function is the Rosenbrock function in ****Appendix Table A.1. This is a unimodal function, whose global optimum is inside a long, narrow, parabolic-shaped flat valley, hence making it difficult to search for the global optimum in many algorithms. The second function is the Griewank function as seen in Appendix Table A.2. It is a multimodal function with multiple local extrema. These local extrema generated by product terms and tall barriers around the global optimum are regularly distributed, and it allows algorithms to sink into local extrema easily (Zhao and Wang, 2016). Hence, these two functions are widely used to evaluate the convergence performance of algorithms.

For the Rosenbrock function, four individuals searching for the global optimum simultaneously start at (−2, −2), (−2, 2), (2, −2), and (2, 2) with 150 iterations. For the Griewank function, six individuals searching for the global optimum simultaneously start at (−6, −6), (−6, 6), (6, −6), (0, −6), (0, 6), and (6, 6) with 150 iterations. The search history, trajectory, and convergence curve are depicted in Fig. 4.7 As shown in the figure, all the history positions of search individuals during iterations are clearly provided, and the distribution of sampled points in the search space can discover whether AEO performs exploration or exploitation. The sparse sampled points show the process of exploring the promising regions and the dense sampled points indicate the process of exploiting a local region. It can be found that the sampled points are sparse in the region further from the global optimum and are dense in the region closer to the global optimum.

Table 4.4 Comparisons of results for composition functions.

Function	Index	AEO	PSO	GA	DE	CS	GSA	ABC
f_{24}	Mean	**2.529E+03**	2.645E+03	2.616E+03	2.615E+03	2.615E+03	2.535E+03	2.617E+03
	Std	1.510E+02	8.508E+00	4.168E−01	2.968E−05	3.295E−03	6.435E+01	4.374E−01
	Best	2.501E+03	2.630E+03	2.616E+03	2.615E+03	2.615E+03	2.506E+03	2.616E+03
f_{25}	Mean	**2.605E+03**	2.631E+03	2.631E+03	2.624E+03	2.633E+03	2.608E+03	2.633E+03
	Std	4.811E+00	5.665E+00	4.335E+00	4.393E+00	3.840E+00	4.734E+00	1.750E+00
	Best	2.600E+03	2.621E+03	2.627E+03	2.602E+03	2.629E+03	2.600E+03	2.629E+03
f_{26}	Mean	**2.701E+03**	2.721E+03	2.714E+03	2.704E+03	2.711E+03	2.702E+03	2.727E+03
	Std	1.429E+00	6.793E+00	3.014E+00	6.924E−01	1.631E+00	3.260E+00	2.056E+00
	Best	2.700E+03	2.712E+03	2.706E+03	2.703E+03	2.709E+03	2.700E+03	2.722E+03
f_{27}	Mean	**2.702E+03**	2.709E+03	2.757E+03	2.704E+03	2.706E+03	2.789E+03	2.705E+03
	Std	1.606E+01	2.500E+01	5.025E+01	1.819E+01	1.793E+01	2.416E+01	1.202E−01
	Best	2.700E+03	2.700E+03	2.700E+03	2.700E+03	2.700E+03	2.722E+03	2.700E+03
f_{28}	Mean	3.655E+03	3.361E+03	3.495E+03	**3.049E+03**	3.122E+03	4.508E+03	3.350E+03
	Std	3.844E+02	2.339E+02	3.693E+02	5.072E+01	5.888E+01	3.007E+02	6.293E+01
	Best	2.908E+03	3.135E+03	3.108E+03	3.000E+03	2.815E+03	3.643E+03	3.186E+03
f_{29}	Mean	7.013E+03	7.051E+03	6.796E+03	**3.663E+03**	3.837E+03	5.546E+03	4.138E+03
	Std	2.217E+03	8.762E+02	8.252E+02	6.744E+01	6.050E+01	1.051E+03	5.010E+01
	Best	3.009E+03	5.422E+03	5.437E+03	3.482E+03	3.734E+03	3.724E+03	4.066E+03
f_{30}	Mean	9.370E+07	1.016E+07	4.107E+03	3.963E+03	7.298E+03	**3.440E+03**	1.600E+05
	Std	5.659E+07	1.528E+07	1.481E+02	2.820E+02	2.115E+03	1.040E+03	8.275E+04
	Best	4.152E+04	3.778E+03	3.903E+03	3.445E+03	5.374E+03	3.100E+03	5.690E+04
f_{31}	Mean	1.198E+06	1.162E+05	1.331E+04	**4.210E+03**	7.445E+03	2.190E+05	2.671E+04
	Std	6.479E+05	7.905E+04	3.082E+03	4.568E+02	7.818E+02	9.727E+04	6.191E+03
	Best	1.992E+05	3.205E+04	8.230E+03	3.533E+03	5.631E+03	7.016E+04	1.724E+04

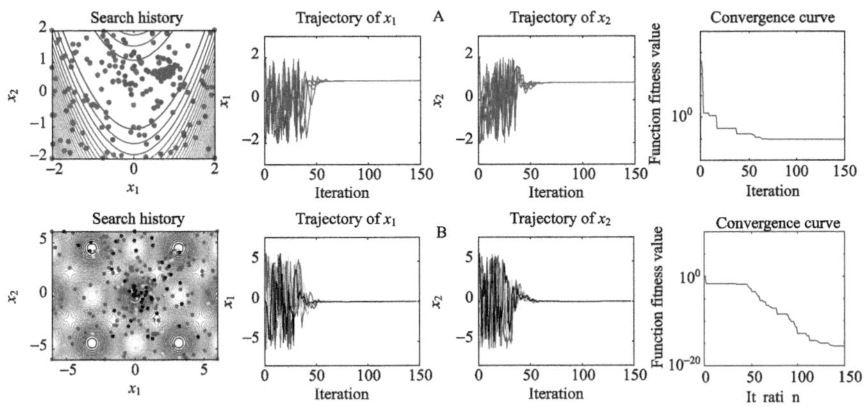

Fig. 4.7 Convergence behaviors of AEO on functions: (A) Rosenbrock function, (B) Griewank function.

The trajectory curves can effectively depict the explorative and exploitative search of AEO, which may trace the positions of all individuals in different dimensions during iterations. In the figure, the trajectory of each variable is depicted. It can be seen that all the trajectories abruptly change in early iterations; this allows AEO to explore the entire search space extensively. Then these abrupt changes gradually reduce to the global optimum in later iterations; these assist AEO to intensively exploit the local space. These trajectories change in the same way as seen in the search history. In addition, we may observe that AEO tends to perform an explorative search before an exploitative search in the trajectory curves.

The convergence curves of all the investigated optimizers for some benchmark functions are shown in Fig. 4.8 to observe their convergence performance. The mean of best-so-far in this figure denotes the mean of the best solutions obtained so far from each algorithm over 30 independent runs. The convergence curves reported in Fig. 4.8 reveal that AEO converges faster than all other optimizers in most benchmark functions and is promising in other cases. It can be concluded that AEO has an excellent convergence capability for these benchmark functions.

4.3.5 Statistical significance analysis

The Wilcoxon signed-rank test (WSRT) can effectively assess the overall performance of algorithms. WSRT is a nonparametric test which can be used to check for statistical significance difference between two algorithms. To determine whether one test outperforms another with certain statistical

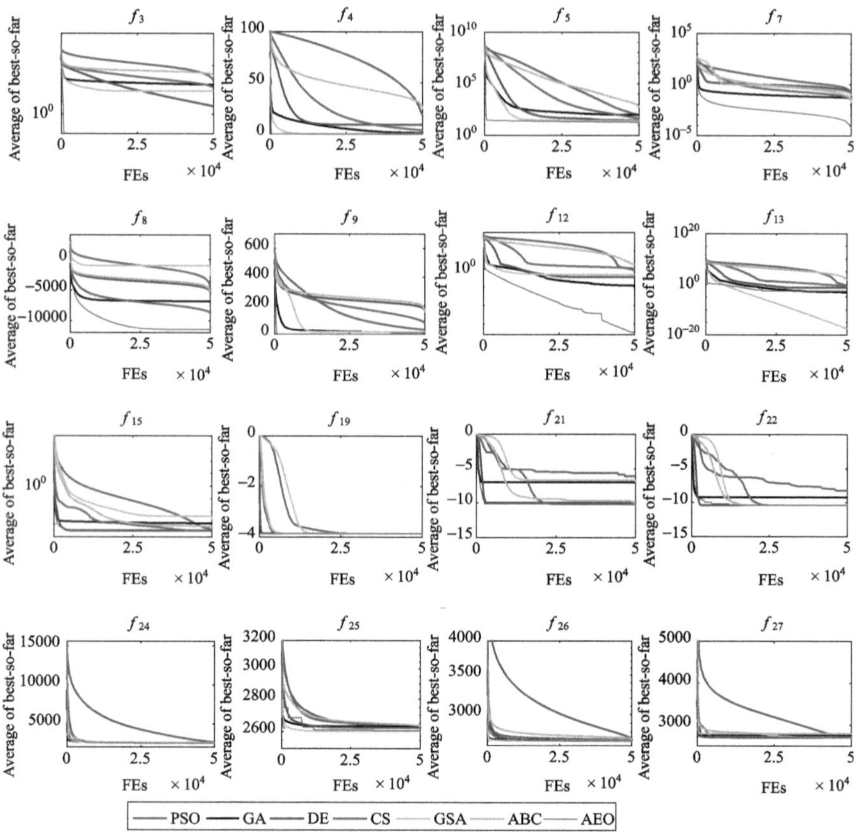

Fig. 4.8 Comparisons of convergence curves for some benchmark functions.

significance, the size of the ranks provided by WSRT is examined. When using WSRT, the R+ and R− related to the comparisons between two algorithms can be calculated and their *P*-values can be obtained. The test results of WSRT on 31 benchmark functions in 30 runs are presented in Tables 4.5 and 4.6, where "=" indicates the case in which there is no statistically significant difference between AEO and the comparative algorithm, "+" indicates the case in which the null hypothesis will be rejected and AEO shows a better performance than the comparative algorithm at 95% significance level ($a = 0.05$), and "−" vice versa. The sum of "+", "−" and "=" for each function in 30 runs is summarized and the corresponding statistical results are listed in Table 4.7. The results demonstrate that compared with its counterparts, AEO shows a better performance at 95% significance level for the majority of functions, although no significant

Table 4.5 Statistical comparisons of WSRT for AEO vs GA, PSO, and DE.

	AEO vs PSO				AEO vs GA				AEO vs DE			
Function	*P*-value	T−	T+	Winner	*P*-value	T−	T+	Winner	*P*-value	T−	T+	Winner
f_1	1.73E−06	0	465	−	1.73E−06	0	465	−	1.73E−06	0	465	−
f_2	1.73E−06	0	465	−	1.73E−06	0	465	−	1.73E−06	0	465	−
f_3	1.73E−06	0	465	−	1.73E−06	0	465	−	1.73E−06	0	465	−
f_4	1.73E−06	0	465	−	1.73E−06	0	465	−	1.73E−06	0	465	−
f_5	3.18E−06	6	459	−	1.92E−06	1	464	−	0.00049	63	402	−
f_6	0.25	0	465	=	1	0	465	=	0.25	0	465	=
f_7	1.73E−06	0	465	−	1.73E−06	0	465	−	1.73E−06	0	465	−
f_8	1.73E−06	0	465	−	1.73E−06	0	465	−	1.73E−06	0	465	−
f_9	1.73E−06	0	465	−	1.73E−06	0	465	−	1.73E−06	0	465	−
f_{10}	1.73E−06	0	465	−	1.73E−06	0	465	−	1.73E−06	0	465	−
f_{11}	1.73E−06	0	465	−	1.73E−06	0	465	−	1.73E−06	0	465	−
f_{12}	1.73E−06	0	465	−	1.73E−06	0	465	−	1.73E−06	0	465	−
f_{13}	2.13E−06	2	463	−	0.057096	140	325	=	0.813017	221	244	=
f_{14}	1	0	465	=	1.22E−05	0	465	−	1	0	465	=
f_{15}	0.000283	56	409	−	2.16E−05	26	439	−	0.001626	341	124	+
f_{16}	1	0	465	=	0.125	0	465	=	1	0	465	=
f_{17}	1	0	465	=	0.003906	0	465	−	1	0	465	=
f_{18}	0.328125	7	458	=	9.80E−05	1	464	−	0.148438	35	430	=
f_{19}	1	0	465	=	3.79E−06	0	465	−	1	0	465	=
f_{20}	0.082328	117	348	=	0.643517	255	210	=	0.175346	114	351	=
f_{21}	9.90E−06	0	465	−	1.73E−06	0	465	−	1	0	465	=
f_{22}	7.06E−05	6	459	−	1.73E−06	0	465	−	0.039063	40	425	−

Continued

Table 4.5 Statistical comparisons of WSRT for AEO vs GA, PSO, and DE—cont'd

	AEO vs PSO				AEO vs GA				AEO vs DE			
Function	***P*-value**	**T−**	**T+**	**Winner**	***P*-value**	**T−**	**T+**	**Winner**	***P*-value**	**T−**	**T+**	**Winner**
f_{23}	0.000237	1.5	463.5	−	1.73E−06	0	465	−	0.5	0	465	=
f_{24}	0.135908	305	160	=	0.023038	122	343	−	0.021827	121	344	−
f_{25}	1.73E−06	0	465	−	1.73E−06	0	465	−	2.13E−06	2	463	−
f_{26}	1.73E−06	0	465	−	1.73E−06	0	465	−	1.49E−05	22	443	−
f_{27}	1.02E−05	18	447	−	0.382034	275	190	=	6.98E−05	168	297	−
f_{28}	0.000664	398	67	+	0.165027	300	165	=	0.07432	305	160	=
f_{29}	0.813017	244	221	=	0.571646	260	205	=	1.36E−05	444	21	+
f_{30}	4.73E−06	455	10	+	1.73E−06	465	0	+	1.73E−06	465	0	+
f_{31}	1.73E−06	465	0	+	1.73E−06	465	0	+	1.73E−06	465	0	+

Table 4.6 Statistical comparisons of WSRT for AEO vs CS, GSA, and ABC.

	AEO vs CS				AEO vs GSA				AEO vs ABC			
Function	***P*-value**	**T−**	**T+**	**Winner**	***P*-value**	**T−**	**T+**	**Winner**	***P*-value**	**T−**	**T+**	**Winner**
f_1	1.73E−06	0	465	−	1.73E−06	0	465	−	1.73E−06	0	465	−
f_2	1.73E−06	0	465	−	1.73E−06	0	465	−	1.73E−06	0	465	−
f_3	1.73E−06	0	465	−	1.73E−06	0	465	−	1.73E−06	0	465	−
f_4	1.73E−06	0	465	−	1.73E−06	0	465	−	1.73E−06	0	465	−
f_5	1.73E−06	0	465	−	1.73E−06	0	465	−	1.73E−06	0	465	−
f_6	1	0	465	=	1	0	465	=	1	0	465	=
f_7	1.73E−06	0	465	−	1.73E−06	0	465	−	1.73E−06	0	465	−
f_8	1.73E−06	0	465	−	1.73E−06	0	465	−	1.73E−06	0	465	−
f_9	1.73E−06	0	465	−	1.72E−06	0	465	−	1.73E−06	0	465	−
f_{10}	1.73E−06	0	465	−	1.73E−06	0	465	−	1.73E−06	0	465	−
f_{11}	1.73E−06	0	465	−	1.73E−06	0	465	−	1.73E−06	0	465	−
f_{12}	1.73E−06	0	465	−	0.571534	205	260	=	1.73E−06	0	465	−
f_{13}	1.73E−06	0	465	−	1.73E−06	465	0	+	1.73E−06	0	465	−
f_{14}	1	0	465	=	2.56E−06	0	465	−	1.73E−06	0	465	−
f_{15}	0.000359	59	406	−	1.73E−06	0	465	−	3.11E−05	30	435	−
f_{16}	1	0	465	=	2.21E−05	0	465	−	0.5	0	465	=
f_{17}	1	0	465	=	1	0	465	=	1.73E−06	0	465	−
f_{18}	0.15332	67	398	=	4.82E−06	0	465	−	1.73E−06	0	465	−
f_{19}	1	0	465	=	2.21E−05	0	465	−	5.99E−07	0	465	−
f_{20}	0.008728	360	105	+	8.72E−05	168	297	−	0.00047	232	233	−
f_{21}	9.57E−07	0	465	−	1.71E−06	0	465	−	1.71E−06	0	465	−
f_{22}	9.25E−06	7	458	−	4.59E−06	0	465	−	1.72E−06	0	465	−

Continued

Table 4.6 Statistical comparisons of WSRT for AEO vs CS, GSA, and ABC—cont'd

Function	AEO vs CS				AEO vs GSA				AEO vs ABC			
	P-value	T−	T+	Winner	*P*-value	T−	T+	Winner	*P*-value	T−	T+	Winner
f_{23}	1.67E−06	0	465	−	7.85E−07	0	465	−	1.73E−06	0	465	−
f_{24}	0.021827	121	344	−	0.093132	278	187	=	0.023038	122	343	−
f_{25}	1.73E−06	0	465	−	9.32E−05	302	163	−	1.73E−06	0	465	−
f_{26}	1.73E−06	0	465	−	0.749871	217	248	=	1.73E−06	0	465	−
f_{27}	0.07387	145	320	=	0.002255	84	381	−	0.07306	205	260	=
f_{28}	0.10891	236	229	=	1.73E−06	0	465	−	0.08319	205	260	=
f_{29}	1.64E−05	442	23	+	0.05667	146	319	=	3.11E−05	435	30	+
f_{30}	1.73E−06	465	0	+	1.73E−06	465	0	+	1.92E−06	464	1	+
f_{31}	1.73E−06	465	0	+	1.73E−06	465	0	+	1.73E−06	465	0	+

Table 4.7 Statistical results of WSRT obtained by AEO.

Function type	AEO vs GA (+/=/−)	AEO vs PSO (+/=/−)	AEO vs DE (+/=/−)	AEO vs CS (+/=/−)	AEO vs GSA (+/=/−)	AEO vs ABC (+/=/−)
Unimodal	6/1/0	6/1/0	6/1/0	6/1/0	6/1/0	6/1/0
Multimodal	6/0/0	5/1/0	5/1/0	6/0/0	4/1/1	6/0/0
Low-dimension	4/6/0	8/2/0	1/8/1	4/5/1	9/1/0	9/1/0
Composition	3/2/3	3/3/2	4/1/3	3/2/3	3/3/2	3/2/3
Total	19/9/3	22/7/2	16/11/4	19/8/4	22/6/3	24/4/3

difference is found between AEO and any other optimizers on function f_6, its mean of best-so-far solution equals the global optimum. From Table 4.7, the statistical analysis shows the fact that AEO considerately outperforms the other optimizers on four typical test functions.

4.3.6 Sensitivity analysis

AEO algorithm has two control parameters: the maximal number of iterations and the size of population. The sensitivity of the control parameters is investigated by changing the value of one parameter while keeping the other one unchanged. Two functions chosen from each of the four types of test functions are used for sensitivity analysis.

The size of population: To explore the effect of the population size on the performance of AEO, the size of population is set as 30, 50, 70, 90, and 100, and the maximal number of iterations is set as 1000. The AEO algorithm is conducted for the benchmark functions under these different population sizes. For each considered function, the AEO algorithm independently runs 30 times for each population size and the results are summarized as the average performance of these runs. Fig. 4.9 and Table 4.8 provide the convergence curves and optimal results of various types of functions with different population sizes. For all the considered functions, the results obtained by AEO are able to offer very high precision for varying population sizes. Varying the population size does not result in a significant change in precision. AEO obtains better convergence towards the global optimum for varying population sizes.

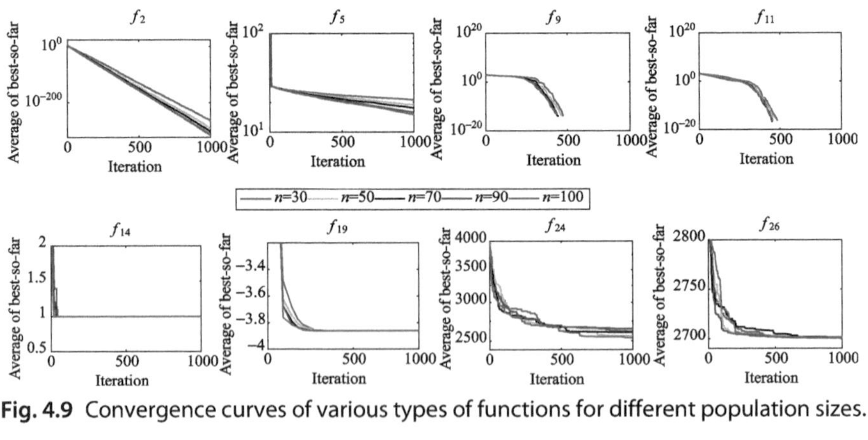

Fig. 4.9 Convergence curves of various types of functions for different population sizes.

Table 4.8 Obtained results of various types of functions for different population sizes.

		Functions							
n	Index	f_2	f_5	f_9	f_{11}	f_{14}	f_{19}	f_{24}	f_{26}
30	Mean	8.251E−266	21.16289	0.000E+00	0.000E+00	0.998004	−3.86278	2620.277	2701.832
	Std	0.000E+00	0.324479	0.000E+00	0.000E+00	0.000E+00	9.367E−16	135.5623	2.175827
50	Mean	7.592E−293	18.94141	0.000E+00	0.000E+00	0.998004	−3.86278	2561.131	2700.384
	Std	0.000E+00	0.829209	0.000E+00	0.000E+00	0.000E+00	9.363E−16	84.18277	0.366186
70	Mean	7.671E−304	17.52624	0.000E+00	0.000E+00	0.998004	−3.86278	2608.156	2700.895
	Std	0.000E+00	0.858311	0.000E+00	0.000E+00	0.000E+00	6.326E−16	133.9838	0.868738
90	Mean	2.322E−315	15.80084	0.000E+00	0.000E+00	0.998004	−3.86278	2651.014	2700.814
	Std	0.000E+00	0.71403	0.000E+00	0.000E+00	0.000E+00	1.431E−16	131.5206	0.936715
100	Mean	1.342E−316	15.10545	0.000E+00	0.000E+00	0.998004	−3.86278	2539.661	2700.93
	Std	0.000E+00	1.071787	0.000E+00	0.000E+00	0.000E+00	5.381E−16	46.11724	0.661825

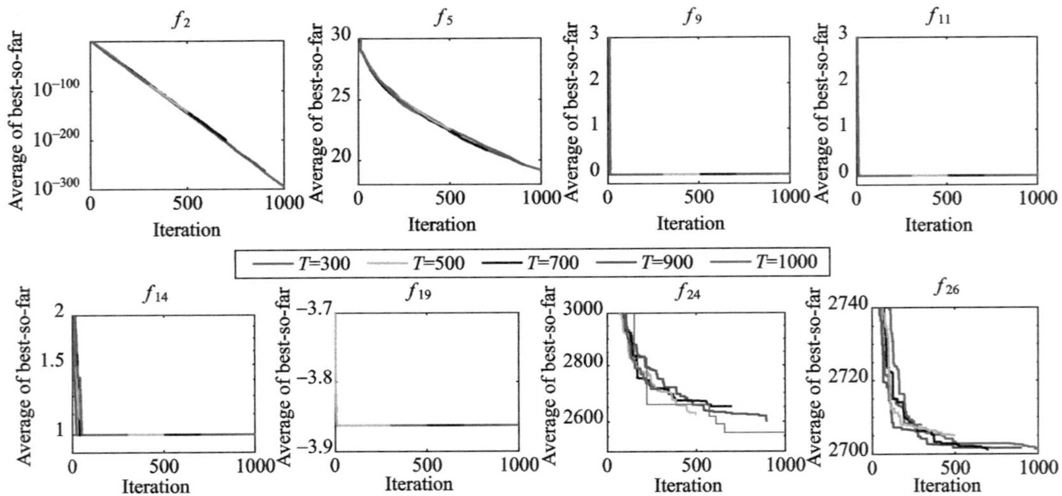

Fig. 4.10 Convergence curves of various types of functions for different number of iterations.

Table 4.9 Obtained results of various types of functions for different number of iterations.

		Functions							
T	Index	f_2	f_5	f_9	f_{11}	f_{14}	f_{19}	f_{24}	f_{26}
300	Mean	4.710E−89	24.49525	0.000E+00	0.000E+00	0.998004	−3.86278	2704.102	2701.304
	Std	5.412E−85	0.515526	0.000E+00	0.000E+00	0.000E+00	1.817E−16	109.782	1.249189
500	Mean	2.827E−142	22.61685	0.000E+00	0.000E+00	0.998004	−3.86278	2611.548	2702.749
	Std	1.903E−148	21.39999	0.000E+00	0.000E+00	0.000E+00	4.71E−16	101.4317	3.695647
700	Mean	2.252E−206	0.443078	0.000E+00	0.000E+00	0.998004	−3.86278	2604.432	2701.462
	Std	9.627E−206	0.858311	0.000E+00	0.000E+00	0.000E+00	3.377E−16	94.99693	1.580398
900	Mean	2.385E−264	20.06242	0.000E+00	0.000E+00	0.998004	−3.86278	2555.139	2701.354
	Std	2.793E−265	0.776597	0.000E+00	0.000E+00	0.000E+00	1.292E−16	55.3923	1.694863
1000	Mean	3.201E−301	18.80743	0.000E+00	0.000E+00	0.998004	−3.86278	2538.878	2700.888
	Std	3.296E−302	0.894924	0.000E+00	0.000E+00	0.000E+00	8.312E−16	62.28167	1.071801

The maximal number of iterations: The maximal number of iterations is set as 300, 500, 700, 900, and 1000, respectively, and the size of population is fixed at 50. Fig. 4.10 and Table 4.9 provide the convergence curves and optimal results of various types of functions for different maximal number of iterations. AEO can converge towards the global optimum as the number of iterations increases. AEO provides good convergence rate for different maximal number of iterations.

4.4 Conclusions

This chapter proposes a new metaheuristic algorithm, named artificial ecosystem-based optimization (AEO), which is motivated from the flow of energy in an ecosystem on the earth. AEO mimics the production, consumption, and decomposition behaviors of living organisms. The AEO is evaluated on 31 mathematical problems, and various analyses such as exploitation, exploration, avoidance of local optima, convergence behavior, statistical significance, quantitative rank and sensitivity, is implemented. The comparative test reveals that the proposed approach is superior to the popular metaheuristics.

References

Abedinpourshotorban, H., Shamsuddin, S.M., Beheshti, D.N.A., Jawawi, Z., et al., 2016. Electromagnetic field optimization: a physics-inspired metaheuristic optimization algorithm. Swarm Evol. Comput. 26, 8–22.

Adeli, H., Cheng, N.T., 1993. Integrated genetic algorithm for optimization of space structures. J. Aerosp. Eng. 6 (4), 315–328.

Akay, B., Karaboga, D., 2009. A modified artificial bee colony algorithm for real-parameter optimization. Inf. Sci. 192, 120–142.

Askarzadeh, A., 2014. Bird mating optimizer:an optimization algorithm inspired by bird mating strategies. Commun. Nonlinear Sci. Numer. Simul. 19 (4), 1213–1228.

Ayman, A.A., 2011. Pid parameters optimization using genetic algorithm technique for electrohydraulic servo control system. Intell. Control. Autom. 2, 888–896.

Bayraktar, Z., Komurcu, M., Bossard, J.A., et al., 2013. The wind driven optimization technique and its application in electromagnetics. IEEE Trans. Antennas Propag. 61 (5), 2745–2757.

Beyer, H.G., Schwefel, H.P., 2002. Evolution strategies—a comprehensive introduction. Nat. Comput. 1 (1), 3–52.

Brown, C., Liebovitch, L.S., Glendon, R., et al., 2007. Lévy flights in Dobe Ju/'hoansi foraging patterns. Hum. Ecol. 35, 129–138.

Chuang, C.L., Jiang, J.A., 2007. Integrated radiation optimization: inspired by the gravitational radiation in the curvature of space-time. In: Evolutionary Computation, CEC 2007, IEEE Congress on. IEEE, pp. 3157–3164.

Civicioglu, P., 2013. Backtracking search optimization algorithm for numerical optimization problems. Appl. Math. Comput. 219 (15), 8121–8144.

Corno, F., Reorda, M.S., Squillero, G., et al., 1998. A new evolutionary algorithm inspired by the selfish gene theory. In: IEEE International Conference on Evolutionary Computation Proceedings. IEEE World Congress on Computational Intelligence IEEE 1976, pp. 575–580.

Cuevas, E., Cienfuegos, M., Zaldívar, D., Pérez-Cisneros, M., et al., 2013. A swarm optimization algorithm inspired in the behavior of the social-spider. Expert Syst. Appl. 40 (16), 6374–6384.

Dorigo, M., Maniezzo, V., Colorni, A., et al., 1996. Ant system: optimization by a colony of cooperating agents. IEEE Trans. Syst. Man Cybern. B 26 (1), 29–41.

Eskandar, H., Sadollah, A., Bahreininejad, A., Hamdi, M., et al., 2012. Water cycle algorithm—a novel metaheuristic optimization method for solving constrained engineering optimization problems. Comput. Struct. 110, 151–166.

Eusuff, M.M., Lansey, K.E., 2003. Optimizing of water distribution network design using the shuffled frog leaping algorithm. J. Water Resour. Plan. Manag. 129 (3), 210–225.

Fathollahi-Fard, A.M., Hajiaghaei-Keshteli, M., Tavakkoli-Moghaddam, R., et al., 2018. The social engineering optimizer (SEO). Eng. Appl. Artif. Intell. 72, 267–293.

Flores, J.J., López, R., Barrera, J., et al., 2011. Gravitational interactions optimization. In: International Conference on Learning and Intelligent Optimization. Springer, Berlin, Heidelberg, pp. 226–237.

Formato, R.A., 2007. Central force optimization: a new metaheuristic with applications in applied electromagnetics. Prog. Electromagn. Res. 77, 425–491.

Gandomi, A.H., Alavi, A.H., 2012. Krill herd:a new bio-inspired optimization algorithm. Commun. Nonlinear Sci. Numer. Simul. 17 (12), 4831–4845.

Geem, Z.W., Kim, J., Loganathan, G.V., et al., 2001. A new heuristic optimization algorithm: harmony search. Trans. Simul. 76 (2), 60–68.

Giannakos, M.N., Krogstie, J., Aalberg, T., et al., 2016. Video-based learning ecosystem to support active learning: application to an introductory computer science course. Smart Learn. Environ. 3 (1), 11.

Hamidreza, R.K., Karim, F., 2008. An improved feature selection method based on ant colony optimization (ACO) evaluated on face recognition system. Appl. Math. Comput. 205, 716–725.

Hare, W., Nutini, J., Tesfamariam, S., et al., 2013. A survey of non-gradient optimization methods in structural engineering. Adv. Eng. Softw. 59, 19–28.

Holland, J.H., 1975. Adaptation in Natural and Artificial Systems. University of Michigan Press, Ann Arbor, MI.

Jain, M., Singh, V., Rani, A., et al., 2019. A novel nature-inspired algorithm for optimization: squirrel search algorithm. Swarm Evol. Comput. 44, 148–175.

Javidy, B., Hatamlou, A., Mirjalili, S., et al., 2015. Ions motion algorithm for solving optimization problems. Appl. Soft Comput. 32, 72–79.

Kaveh, A., Bakhshpoori, T., 2016. Water evaporation optimization: a novel physically inspired optimization algorithm. Comput. Struct. 167, 69–85.

Kaveh, A., Dadras, A., 2017. A novel meta-heuristic optimization algorithm: thermal exchange optimization. Adv. Eng. Softw. 110, 69–84.

Kaveh, A., Farhoudi, N., 2013. A new optimization method: dolphin echolocation. Adv. Eng. Softw. 59, 53–70.

Kaveh, A., Talatahari, S., 2010. A novel heuristic optimization method: charged system search. Acta Mech. 213 (3), 267–289.

Kennedy, J., Eberhart, R., 1995. Particle swarm optimization. In: Proceedings of Sixth International Symposium on Micro Machine and Human Science, pp. 39–43.

Kiran, M.S., 2015. TSA: tree-seed algorithm for continuous optimization. Expert Syst. Appl. 42 (19), 6686–6698.

Kirkpatrick, S., Gelatt, C.D., Vecchi, M.P., et al., 1983. Optimization by simulated annealing. Science 220 (4598), 671–680.
Krause, J., Cordeiro, J., Parpinelli, R.S., Lopes, H.S., et al., 2013. A survey of swarm algorithms applied to discrete optimization problems. In: Yang, X.S., Cui, Z.H., Xiao, R.B., Gandomi, A.H., Karamanoglu, M. (Eds.), Swarm Intelligence and Bio-Inspired Computation: Theory and Applications. Elsevier Science & Technology Books, Elsevier, London, pp. 169–191.
Kripka, M., Kripka, R.M.L., 2008. Big crunch optimization method. In: International Conference on Engineering Optimization, Brazil, pp. 1–5.
Lai, C., Shao, Q., Chen, X., Wang, Z., 2016. Flood risk zoning using a rule mining based on ant colony algorithm. J. Hydrol. 542, 268–280.
Li, M.D., Zhao, H., Weng, X.W., et al., 2016. A novel nature-inspired algorithm for optimization: virus colony search. Adv. Eng. Softw. 92, 65–88.
Liang, J.J., Qu, B.Y., Suganthan, P.N., et al., 2013. Problem Definitions and Evaluation Criteria for the CEC 2014 Special Session and Competition on Single Objective Real-Parameter Numerical Optimization. Computational Intelligence Laboratory, Zhengzhou University, Zhengzhou China and Technical Report, Nanyang Technological University, Singapore.
Milner, S., Davis, C., Zhang, H., Llorca, J., et al., 2012. Nature-inspired self-organization, control, and optimization in heterogeneous wireless networks. IEEE Trans. Mob. Comput. 11 (7), 1207–1222.
Mirjalili, S.A., Hashim, S.Z.M., 2012. BMOA: binary magnetic optimization algorithm. Int. J. Mach. Learn. Comput. 2 (3), 204.
Mirjalili, S., Lewis, A., 2016. The whale optimization algorithm. Adv. Eng. Softw. 95, 51–67.
Mirjalili, S., Gandomi, A.H., Mirjalili, S.Z., et al., 2017. Salp swarm algorithm: a bio-inspired optimizer for engineering design problems. Adv. Eng. Softw. 114, 163–191.
Moghaddam, F.F., Moghaddam, R.F., Cheriet, M., et al., 2012. Curved space optimization: a random search based on general relativity theory. arXiv 1208, 2214v. preprint arXiv.
Mühlenbein, H., Gorges-Schleuter, M., Krämer, O., et al., 1988. Evolution algorithms in combinatorial optimization. Parallel Comput. 7 (1), 65–85.
Nayak, B., Misra, B., Choudhury, T.R., et al., 2018. Meta-heuristic optimization algorithms for design of gain constrained state variable filter. Int. J. Electron. Commun. 93, 7–18.
Oftadeh, R., Mahjoob, M.J., Shariatpanahi, M., et al., 2010. A novel meta-heuristic optimization algorithm inspired by group hunting of animals: hunting search. Comput. Math. Appl. 60 (7), 2087–2098.
O'Neill, R.V., Deangelis, D.L., Waide, J.B., et al., 1986. A Hierarchical Concept of Ecosystems (No. 23). Princeton University Press, New Jersey.
Pan, W.T., 2012. A new fruit fly optimization algorithm: taking the financial distress model as an example. Knowl.-Based Syst. 26, 69–74.
Passino, K.M., 2002. Biomimicry of bacterial foraging for distributed optimization and control. IEEE Control. Syst. 22 (3), 52–67.
Patel, V.K., Savsani, V.J., 2015. Heat transfer search (HTS): a novel optimization algorithm. Inf. Sci. 324, 217–246.
Rao, R.V., Savsani, V.J., Vakharia, D.P., et al., 2012. Teaching-learning-based optimization: an optimization method for continuous non-linear large scale problems. Inf. Sci. 183 (1), 1–15.
Rashedi, E., Nezamabadi-Pour, H., Saryazdi, S., et al., 2009. GSA: a gravitational search algorithm. Inf. Sci. 179 (13), 2232–2248.
Saremi, S., Mirjalili, S., Lewis, A., et al., 2017. Grasshopper optimisation algorithm: theory and application. Adv. Eng. Softw. 105, 30–47.
Shah-Hosseini, H., 2009. The intelligent water drops algorithm: a nature-inspired swarm-based optimization algorithm. Int. J. Bio-Inspir. Comput. 1 (1–2), 71–79.

Shah-Hosseini, H., 2011. Principal components analysis by the galaxy-based search algorithm: a novel metaheuristic for continuous optimization. Int. J. Comput. Sci. Eng. 6 (1–2), 132–140.

Shen, J., Li, Y., 2009. Light ray optimization and its parameter analysis. In: Computational Sciences and Optimization, CSO 2009, International Joint Conference on. IEEE, 2, pp. 918–922.

Simon, D., 2009. Biogeography-based optimization. IEEE Trans. Evol. Comput. 12 (6), 702–713.

Storn, R., Price, K., 1997. Differential evolution—a simple and efficient heuristic for global optimization over continuous spaces. J. Glob. Optim. 11, 341–359.

Talbi, E.G., 2009. Metaheuristics: From Design to Implementation. vol. 74 John Wiley & Sons, New York.

Tamura, K., Yasuda, K., 2011. Primary study of spiral dynamics inspired optimization. IEEE Trans. Electr. Electron. Mater. 6 (S1), 98–100.

Tarek, H., Mohamed, S., Moustafa, K., et al., 2011. Incorporating rework into construction schedule analysis. Autom. Constr. 20, 1051–1059.

Uymaz, S.A., Tezel, G., Yel, E., et al., 2015. Artificial algae algorithm (AAA) for nonlinear global optimization. Appl. Soft Comput. 31, 153–171.

Vikhar, P., 2016. Evolutionary algorithm: a classical search and optimization technique. Int. J. Pure Appl. Res. Eng. Technol. 4 (9), 758–766.

Viswanathan, G.M., Afanasyev, V., Buldyrev, S.V., et al., 1996. Lévy flight search patterns of wandering albatrosses. Nature 381, 413–415.

Wang, L., Zhao, W., Tian, Y., et al., 2018. A bare bones bacterial foraging optimization algorithm. Cogn. Syst. Res. 52, 301–311.

Xie, L., Zeng, J., 2010. The performance analysis of artificial physics optimization algorithm driven by different virtual forces. ICIC Express Lett. 4 (1), 239–244.

Yang, X.S., 2010. Firefly algorithm, stochastic test functions and design optimization. Int. J. Bio-Inspir. Comput. 2 (2), 78–84.

Yang, X.S., Deb, S., 2009. Cuckoo search via Lévy flights. In: Nature & Biologically Inspired Computing, NaBIC 2009, World Congress on IEEE, pp. 210–214.

Yang, X.S., Deb, S., 2013. Multiobjective cuckoo search for design optimization. Comput. Oper. Res. 40, 1616–1624.

Zarand, G., Pazmandi, F., Pál, K.F., et al., 2002. Using hysteresis for optimization. Phys. Rev. Lett. 89 (15), 150–201.

Zhang, H., Cao, X., Ho, J.K., et al., 2016. Object-level video advertising: an optimization framework. IEEE Trans. Ind. Inf. 13 (2), 520–531.

Zhao, W., Wang, L., 2016. An effective bacterial foraging optimizer for global optimization. Inf. Sci. 329, 719–735.

Zhao, W., Wang, L., Zhang, Z., 2019a. A novel atom search optimization for dispersion coefficient estimation in groundwater. Future Gener. Comput. Syst. 91, 601–610.

Zhao, W., Wang, L., Zhang, Z., 2019b. Artificial ecosystem-based optimization: a novel nature-inspired meta-heuristic algorithm. Neural Comput. & Applic. https://doi.org/10.1007/s00521-019-04452-x.

Zhao, W., Wang, L., Zhang, Z., 2019c. Supply-demand-based optimization: a novel economics-inspired algorithm for global optimization. IEEE Access 7, 73182–73206.

Zheng, Y.J., 2015. Water wave optimization: a new nature-inspired metaheuristic. Comput. Oper. Res. 5, 1–11.

Zheng, M., Liu, G., Zhou, C., et al., 2010. Gravitation field algorithm and its application in gene cluster. Algorithms Mol. Biol. 5 (1), 32.

CHAPTER 5

Engineering applications of artificial ecosystem-based optimization

Contents

To further test the ability of the artificial ecosystem-based optimization (AEO) algorithm to handle real-world problems, an extensive study using AEO has been conducted on some real-world engineering problems.

5.1 Engineering optimization using the AEO algorithm

Five constrained engineering design problems, including tension/compression spring design, pressure vessel design, welded beam design, speed reducer design, and multiple disc clutch brake design, are utilized to the validity of this proposed algorithm. However, these problems are all constraint optimization problems that involve inequality and equality constraints. For simplicity, penalty functions are used to handle these

New Optimization Algorithms and their Applications
https://doi.org/10.1016/B978-0-323-90941-9.00005-3

constraint problems in AEO. When using penalty functions, every search individual that violates a constraint at any level is assigned a big function fitness value. In this way, these constrained problems can be converted into unconstrained problems by introducing penalty functions. This method is very effective and is easy to implement for AEO without the need for modification of the algorithm. AEO does not need to adjust any other parameters except for the size of population and the maximal number of function evaluations, which indicates the consistence of AEO performance among different problems.

5.1.1 Tension/compression spring design

This problem (Belegundu, 1982) needs to minimize the weight of a tension/compression spring and meet the constraint requirements on shear stress, deflection, and surge frequency as shown in Fig. 5.1. There are three decision variables: wire diameter (d), mean coil diameter (D), and number of active coils (N). The optimization equation for this problem is given in Appendix B.1.

The comparisons of statistical results for AEO with the reported methods in the literature, including GA2, GA3, CA, CPSO, HPSO, PSO2, QPSO, UPSO, CDE, SSB, and ($\mu + \lambda$)ES, are provided in Table 5.1. In Table 5.1, the two sets of statistical results are obtained using AEO for 2000 FEs and 25,000 FEs, respectively. As seen in the table, compared with the other algorithms, AEO under predefined maximum function evaluations is very promising. More specifically, for the second set of statistical results, AEO is far superior to all the other metaheuristic methods in terms of four different indices using a smaller number of FEs. The comparisons of best solutions offered by investigated optimizers are presented in Table 5.2.

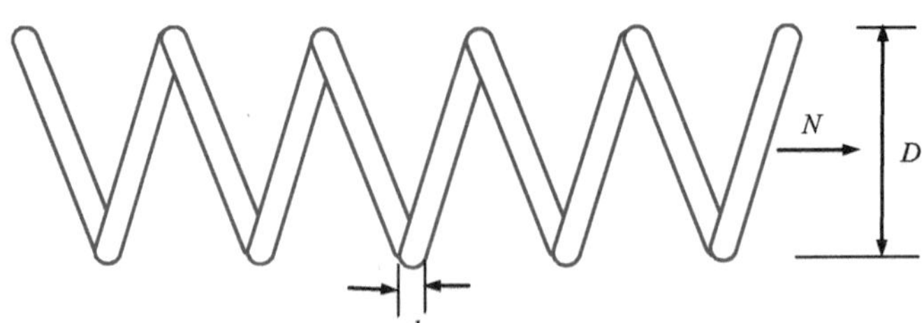

Fig. 5.1 Tension/compression string design problem.

Table 5.1 Comparisons of statistical results using reported optimizers in literature for tension/compression spring design.

Methods	Worst	Mean	Best	Std	FEs
GA2 (Coello, 2000)	0.0128221	0.0127690	0.0127047	3.9390E−05	900,000
GA3 (Coello and Montes, 2002)	0.0129730	0.0127420	0.0126810	5.9000E−05	80,000
CA (Coello and Becerra, 2004)	0.0151156	0.0135681	0.0127210	8.4215E−04	50,000
CPSO (He and Wang, 2006)	0.0129240	0.0127300	0.0126747	5.1985E−04	200,000
HPSO (He and Wang, 2007)	0.0127191	0.0127072	0.0126652	1.5824E−05	75,000
PSO2 (Dos Santos Coelho, 2010)	0.0718020	0.0195550	0.0128570	0.0116620	2000
QPSO (Dos Santos Coelho, 2010)	0.0181270	0.0138540	0.0126690	1.3410E−03	2000
UPSO (Parsopoulos and Vrahatis, 2005)	0.0503651	0.0229478	0.0131200	7.2057E−03	10,000
CDE (Huang et al., 2007)	0.0127900	0.0127030	0.0126702	2.7000E−05	240,000
SSB (Ray and Liew, 2003)	0.0167173	0.0129227	0.0126692	5.9000E−04	25,167
$(\mu + \lambda)$ES (Mezura-Montes and Coello, 2005)	NA	0.0131650	0.0126890	3.9000E−04	30,000
AEO	0.0142060	0.0133186	0.0127324	3.4611E−04	2000
AEO	0.0127271	0.01272712	0.0126662	2.5401E−05	25,000

Table 5.2 Comparisons of best solutions offered by reported optimizers for tension/compression spring design.

	GA3	CA	CPSO	HPSO	CDE	$(\mu + \lambda)$ES	AEO
$x_1(d)$	0.051989	0.050000	0.051728	0.051706	0.051609	0.052836	0.051897
$x_2(D)$	0.363965	0.317395	0.357644	0.357126	0.354714	0.384942	0.361751
$x_3(N)$	10.890522	14.031795	11.244543	11.265083	11.410831	9.807729	10.879842
g_1	−0.000013	0	−8.25E−04	−3.06E−06	−0.000039	−0.000001	−1.26E−05
g_2	−0.000021	−0.000075	−2.52E−05	−1.39E−06	−0.000183	0	−6.75E−08
g_3	−4.061338	−3.967960	−4.051306	−4.054583	−4.048627	−4.106146	−4.063548
g_4	−0.722698	−0.755070	−0.727085	−0.727445	−0.729118	−0.708148	−0.72427
f_1	0.0126810	0.0127210	0.0126747	0.0126652	0.0126702	0.012689	0.0126662

Fig. 5.2 illustrates the convergence curve and each constraint value with respect to FEs, respectively. As observed in this figure, the individuals obtain the infeasible solutions violating two constraints in their early iterations, resulting in a higher function value. With the increase of iterations, the individuals quickly obtain feasible solutions satisfying all the constraints, thus rapidly decreasing the function value. Obviously, AEO performs a good convergence rate because it tends to reach the best solution in almost 500 iterations. This may be due to a special search mechanism and constraint handling strategy of AEO in which it initially explores the entire problem space and then rapidly approaches the optimal solution (Zhao et al., 2019).

5.1.2 Pressure vessel design

This case (Kannan and Kramer, 1994) needs to minimize the fabrication cost of a cylindrical pressure vessel and meet the constraint requirements for buckling load end deflection of the beam, shear stress, and bending stress in the beam. A cylindrical vessel is capped at both ends by hemispherical heads as depicted in Fig. 5.3. This problem includes four decision variables: thickness of weld (T_s), length of the clamped bar (T_h), height of the bar (R), and thickness of the bar (L). The optimization equation for this problem is given in Appendix B.2.

This problem is handled using many reported metaheuristic approaches, such as GA2, GA3, CPSO, HPSO, PSO-DE, PSO2, CDE, ABC, QPSO, ($\mu + \lambda$)ES, and CSA. Table 5.3 shows the statistical results obtained from these approaches with those found by the proposed one. The two sets of statistical results are obtained using AEO for 8000 FEs and 30,000 FEs,

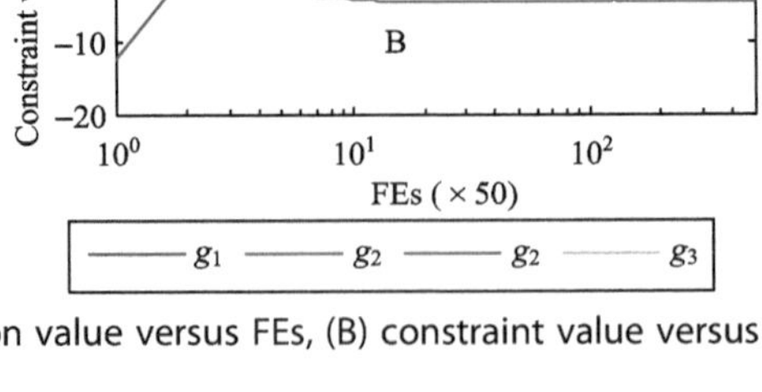

Fig. 5.2 (A) Function value versus FEs, (B) constraint value versus FEs.

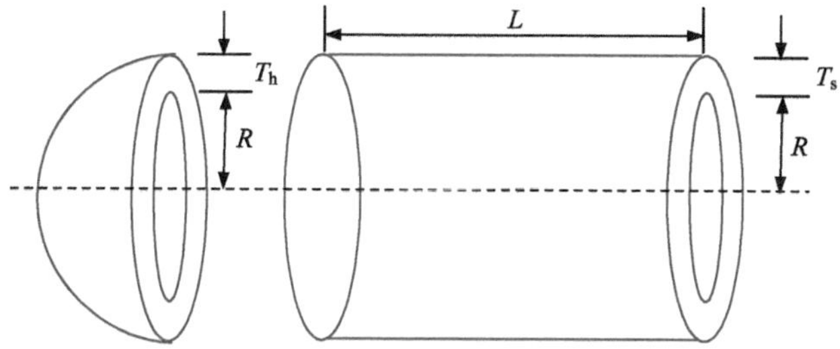

Fig. 5.3 Pressure vessel design problem.

respectively. From this table, for the first set of results with the same FEs, the results of AEO are superior to those of PSO2 and QPSO in terms of four different indices. For the second set of results, in terms of the "Best" index, with a smaller number of FEs, the results offered by AEO outperform the results offered by all the other algorithms. Meanwhile, in terms of the "Worst" and "Mean" indices, the performance of AEO is promising. The comparisons of best solutions offered by reported optimizers are represented in Table 5.4.

Fig. 5.4 illustrates the convergence curve and each constraint value with respect to FEs, respectively. As shown in this figure, during the initial stage of iterations, the solutions of individuals are in the infeasible region because the solutions violate the third constraint. As 50 FEs are completed, the infeasible solutions are transformed into the feasible ones that are likely to converge to near the global optimum in about 2000 FEs. This characteristic, observed in other engineering problems, can be viewed as a merit of the proposed approach (Zhao et al., 2019).

5.1.3 Welded beam design

In this problem (Coello, 2000), the fabrication cost of a welded beam is to be minimized subject to constraints on shear stress, bending stress in the beam, buckling load on the bar, end deflection of the beam, and side constraints. There are four decision variables as shown in Fig. 5.5, including thickness of weld (h), length of attached part of bar (l), height of the bar (t), and thickness of the bar (b). The optimization equation for this problem is given in Appendix B.3.

The statistical results of some competitive metaheuristic methods, including GA2, GA3, CPSO, HPSO, PSO-DE, WOA, EPSO, ABC, ($\mu + \lambda$)ES, and SC, are compared with the proposed approach, and their comparative results are listed in Table 5.5. In this table, two sets of statistical

Table 5.3 Comparisons of statistical results using reported optimizers in literature for pressure vessel design.

Methods	Worst	Mean	Best	Std	FEs
GA2	6308.4970	6293.8432	6288.7445	7.4133	900,000
GA3	6469.3220	6177.2533	6059.9463	130.9297	80,000
CPSO	6363.8041	6147.1332	6061.0777	86.4500	240,000
HPSO	6288.6770	6099.9323	6059.7143	86.2000	81,000
PSO-DE	NA	6059.7140	6059.714	NA	42,100
PSO2	14,076.3240	8756.6803	6693.7212	1492.5670	8000
QPSO	8017.2816	6440.3786	6059.7209	479.2671	8000
CDE	6371.0455	6085.2303	6059.7340	43.0130	204,800
ABC (Akay and Karaboga, 2012)	NA	6245.3080	6059.714339	205.0000	30,000
$(\mu + \lambda)$ES	6820.397461	6379.938037	6059.701610	210.0000	30,000
CSA (Askarzadeh, 2016)	7332.8410	6342.4990	6059.7140	384.9450	250,000
AEO	7110.1964	6615.7111	6030.0039	216.7381	8000
AEO	6820.8007	6136.3019	59,945.0695	161.2901	30,000

Table 5.4 Comparisons of best solutions offered by reported optimizers for pressure vessel design.

	GA3	CPSO	HPSO	CDE	ABC	$(\mu + \lambda)$ES	AEO
$x_1(T_s)$	0.812500	0.812500	0.8125	0.812500	0.8125	0.8125	0.8374205
$x_2(T_h)$	0.437500	0.437500	0.4375	0.437500	0.4375	0.4375	0.413937
$x_3(R)$	42.097398	42.091266	42.0984	42.098411	42.098446	42.098446	43.389597
$x_4(L)$	176.654047	176.746500	176.6366	176.637690	176.636596	76.636596	161.268592
g_1	−0.000020	−0.000139	−8.80E−07	−6.677E−07	0	0	−1.3666E−6
g_2	−0.035891	−0.035949	−3.58E−02	−0.035881	−0.035881	−0.035880	5.3169E−7
g_3	−27.886075	−116.382700	−3.12260	−3.683016	−0.000226	0	−2.641622
g_4	−63.345953	−63.253500	−63.36340	−63.36231	−63.363404	63.363404	−78.731407
f_2	6059.946341	6061.0777	6059.7143	6059.7340	6059.714339	6059.7143	59,945.0695

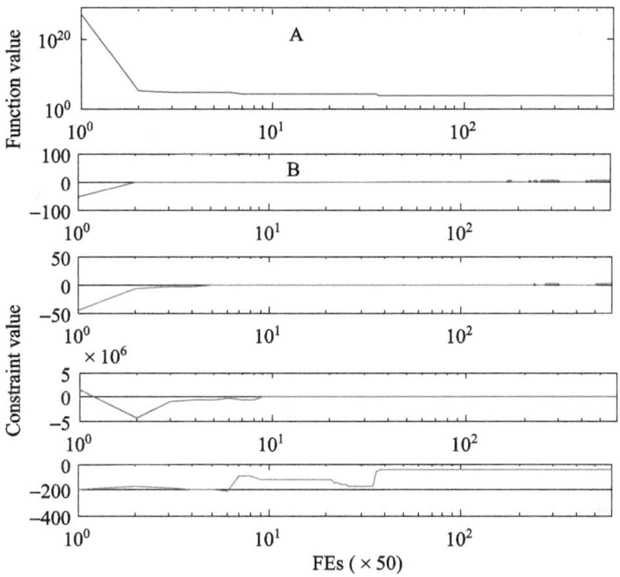

Fig. 5.4 (A) Function value versus FEs, (B) constraint value versus FEs.

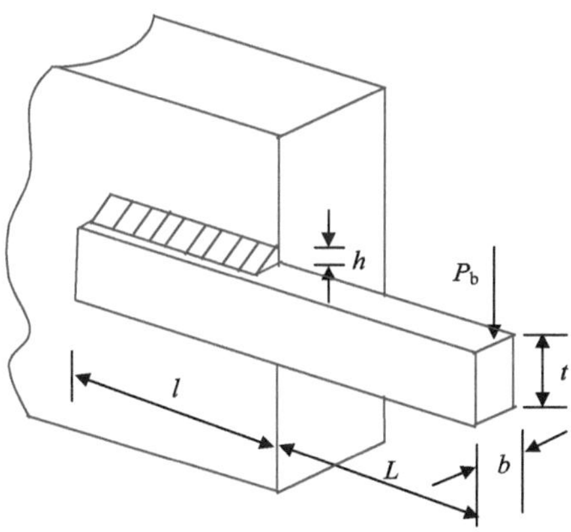

Fig. 5.5 Pressure vessel design problem.

results found by AEO with 9900 FEs and 15,000 FEs are provided, respectively. As the table shows, in the first set of results with the same FEs, AEO performs better than WOA in terms of the "Mean" and "Std" indices. For the second set of results, AEO provides more promising results in terms of four indices using a much smaller number of FEs than those provided by all the other methods except PSO-DE. However, the results obtained by AEO are slightly inferior to the results given by PSO-DE in terms of the "Worst"

Table 5.5 Comparisons of statistical results using reported optimizers in literature for welded beam design.

Methods	Worst	Mean	Best	Std	FEs
GA2	1.7858350	1.7719730	1.7483090	1.1200E−02	900,000
GA3	1.9934080	1.7926540	1.7282260	7.4700E−02	80,000
CPSO	1.7821430	1.7488310	1.7280240	1.2900E−02	240,000
HPSO	1.8142950	1.7490400	1.7248520	4.0100E−02	81,000
PSO-DE	1.7248520	1.7248520	1.7248520	6.7000E−16	66,600
CDE	1.8241050	1.7681580	1.7334620	2.2194E−02	204,800
WOA (Mirjalili and Lewis, 2016)	NA	1.7320000	NA	0.0226000	9900
EPSO (Ngo et al., 2016)	1.7472200	1.7282190	1.7248530	5.6200E−03	50,000
ABC	NA	1.7419130	1.7248520	3.100E−02	30,000
$(\mu + \lambda)$ES	NA	1.7776920	1.7248520	8.800E−02	30,000
SC	6.3996780	3.0025883	2.3854347	9.600E−01	33,095
AEO	1.7383026	1.7274068	1.7248686	4.39E−03	9900
AEO	1.7255664	1.7250057	1.7248520	2.4763E−04	15,000

and "Mean" indices. The comparisons of best solutions offered by investigated metaheuristic methods are represented in Table 5.6. Fig. 5.6 depicts the function value and each constraint value with respect to FEs for the welded beam design problem (Zhao et al., 2019).

5.1.4 Speed reducer design

In this problem (Mezura-Montes and Coello, 2005), the weight of the speed reducer is to be minimized subject to constraints on the bending stress of the gear teeth, surface stress, transverse deflections of the shafts, and stresses in the shafts. There are seven decision variables as shown in Fig. 5.7: face width (b), module of teeth (m), number of teeth in the pinion (z), length of the first shaft between bearings (l_1), length of the second shaft between bearings (l_2), and diameter of first (d_1) and second shafts (d_2). The optimization equation for this problem is given in Appendix B.4.

The results of the proposed algorithm are compared with those of the other considered algorithms in terms of different statistical indexes. These reported methods include SC, DELC, HEAA, PSO-DE, DEDS, ABC, ($\mu + \lambda$)ES, and MDE. The comparisons of statistical results are presented in Table 5.7. As this table shows, the results offered by AEO are as competitive as those by DELC and DEDS in terms of the "Worst", "Mean" and "Best" indices, which outperform the other considered algorithms except ABC. However, AEO can reach the optimum solution much faster with considerably less computational efforts than all other reported methods. The comparisons of best solutions offered by reported optimizers are given in Table 5.8. Fig. 5.8 shows the function value and each constraint value with respect to FEs for speed reducer design problem (Zhao et al., 2019).

5.1.5 Multiple disc clutch brake design

This problem (Osyczka, 2002), as shown in Fig. 5.9, needs to minimize the mass of the multiple disc clutch brake using five discrete decision variables: inner radius ($r_i = 60, 61, 62, \cdots, 80$), outer radius ($r_o = 90, 91, 92, \cdots, 110$), thickness of the disc ($t = 1, 1.5, 2, 2.5, 3$), actuating force ($F = 600, 610, 620, \cdots, 1000$), and number of friction surfaces ($Z = 2, 3, 4, 5, 6, 7, 8, 9$). The optimization equation for this problem is given in Appendix B.5.

The comparisons of statistical results for the reported optimizers including TLBO and ABC are shown in Table 5.9. As this table shows, in terms of all the four indices, with less computational efforts, the developed algorithm offers its best results of all the considered optimizers for this case. The comparisons of best solutions offered by reported optimizers are also represented

Table 5.6 Comparisons of best solutions offered by reported optimizers for welded beam design.

	GA3	CPSO	HPSO	CDE	ABC	$(\mu + \lambda)$ES	AEO
$x_1(T_s)$	0.205986	0.202369	0.20573	0.203137	0.205730	0.205730	0.2057296
$x_2(T_h)$	3.471328	3.544214	3.470489	3.542998	3.470489	3.470489	3.4704886
$x_3(R)$	9.020224	9.048210	9.036624	9.033498	9.036624	9.036624	9.0366239
$x_4(L)$	0.206480	0.205723	0.20573	0.206179	0.205730	0.205729	0.2057296
g_1	−0.074092	−12.839796	−0.025399	−44.578568	0	0	−1.82E−11
g_2	−0.266227	−1.247467	−0.053122	−44.663534	0.000002	0.000002	−2.18E−11
g_3	−0.000495	−0.001498	0	−0.003042	0	0	−1.23E−12
g_4	−3.430043	−3.429347	−3.432981	−3.423726	−3.432984	−3.432984	−3.4329837
g_5	−0.080986	−0.079381	−0.08073	−0.078137	−0.08073	−0.08073	−0.0807296
g_6	−0.235514	−0.235536	−0.235540	−0.235557	−0.235540	−0.235540	−0.2355403
g_7	−58.666440	−11.681355	−0.031555	−38.028268	0.000001	0.000001	0
f_3	1.728226	1.728024	1.724852	1.733462	1.724852	1.724852	1.7248520

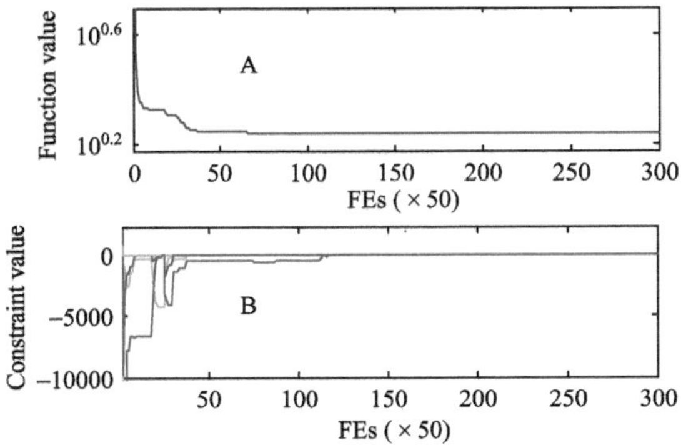

Fig. 5.6 (A) Function value versus FEs, (B) constraint value versus FEs.

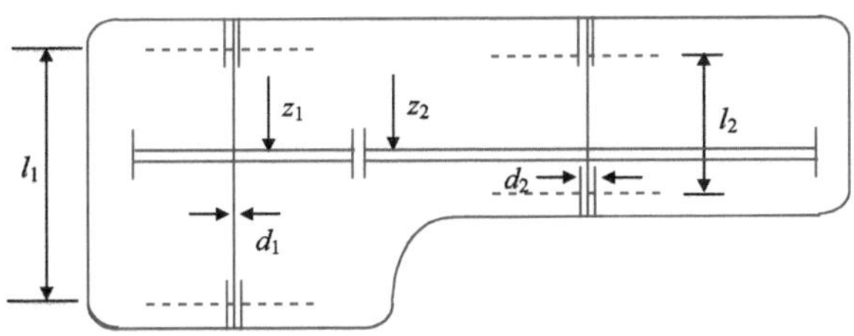

Fig. 5.7 Speed reducer design.

in Table 5.10. Fig. 5.10 shows the convergence curves for three algorithms. As observed in Fig. 5.10A, the convergence rate of TLBO is faster than ABC in the early iterations, but with the increase of iterations, the convergence of both algorithms becomes almost the same (Rao et al., 2011). From Fig. 5.10B, AEO provides a faster convergence rate than the two optimizers that offer the best solution after about 600 FEs. However, our approach can find the same solution only using 500 FEs (Zhao et al., 2019).

These overall tests suggest that the proposed AEO can be viewed as a potential alternative for solving various optimization problems. It can be seen that the comparisons on five classical constrained engineering problems demonstrate that AEO may offer a faster convergence rate and high-quality solutions with less computational efforts.

5.2 Static economic load dispatch problem

5.2.1 Problem description

In order to ensure the economical operation of power systems, it is necessary to carry out the economic dispatch of the power system load, that is, allocating the load of power systems reasonably among the units can reduce the

Table 5.7 Comparisons of statistical results using reported optimizers in literature for speed reducer design.

Methods	Worst	Mean	Best	Std	FEs
SC	3009.9647360	3001.7582640	2994.7442410	4	54,456
DELC (Wang and Li, 2010)	2994.4710660	2994.4710660	2994.4710660	1.9000E − 12	30,000
HEAA	2994.7523110	2994.6133680	2994.4991070	7.0000E − 02	40,000
PSO-DE	2996.3482040	2996.3481740	2996.3481670	6.4000E − 06	54,350
DEDS	2994.4710660	2994.4710660	2994.4710660	3.6000E − 12	30,000
ABC	NA	2997.0580000	2997.0580000	0	30,000
$(\mu + \lambda)$ES	NA	2996.3480940	2996.3480940	0	30,000
MDE (Montes et al., 2006)	NA	2996.3672200	2996.3566890	8.2000E − 03	24,000
AEO	2994.4710667	2994.4710662	2994.4710661	1.2391E − 07	22,000

Table 5.8 Comparisons of best solutions offered by reported optimizers for speed reducer design.

	SC	HEAA	DEDS	ABC	$(\mu+\lambda)$ES	MDE	AEO
$x_1(b)$	3.50000	3.5000228993	3.5000000000	3.499999	3.499999	3.500010	3.5
$x_2(m)$	0.70000	0.7000003924	0.7000000000	0.7	0.699999	0.700000	0.7
$x_3(z)$	17	17.0000128592	17	17	17	17	17
$x_4(l_1)$	7.327602	7.3004277414	7.3000000000	7.3	7.300000	7.300156	7.3
$x_5(l_2)$	7.715321	7.7153774494	7.7153199115	7.8	7.800000	7.800027	7.7153199
$x_6(d_1)$	3.350267	3.3502309666	3.3502146661	3.350215	3.350215	3.350221	3.3502146
$x_7(d_2)$	5.286655	5.2866636970	5.2866544650	5.287800	5.286683	5.286685	5.2866545
g_1	−0.073915	NA	NA	−0.073915	−0.073915	−0.073918	−0.0739152
g_2	−0.197999	NA	NA	−0.197999	−0.197998	−0.198001	−0.1979985
g_3	−0.493501	NA	NA	−0.499172	−0.499172	−0.499144	−0.4991722
g_4	−0.904644	NA	NA	−0.901555	−0.901472	−0.901471	−0.9046439
g_5	0	NA	NA	0	0	−0.000005	1.4322E − 13
g_6	0.000633	NA	NA	0	0	−0.000001	5.9175E − 13
g_7	−0.7025	NA	NA	−0.702500	−0.702500	−0.702500	−0.7025
g_8	0	NA	NA	0	0	−0.000003	6.8002E − 13
g_9	−0.583333	NA	NA	−0.583333	−0.583333	−0.583332	−0.7958333
g_{10}	−0.054889	NA	NA	−0.051326	−0.051325	−0.051345	−0.0513257
g_{11}	0	NA	NA	−0.010695	−0.010852	−0.010856	9.0150E − 14
f_4	2994.744241	2994.499107	2994.471066	2997.058412	2996.348094	2996.356689	2994.4710661

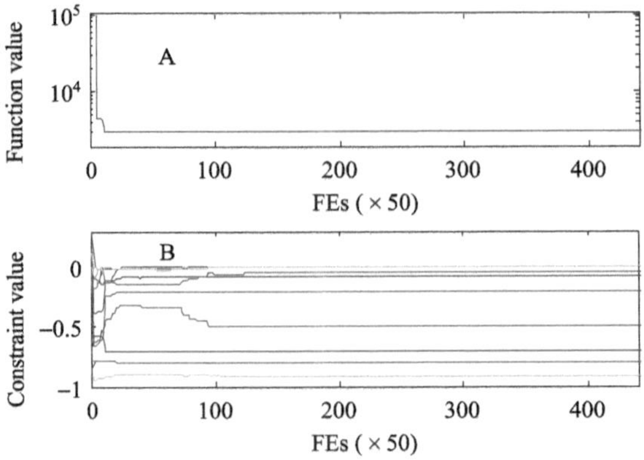

Fig. 5.8 (A) Function value versus FEs, (B) constraint value versus FEs.

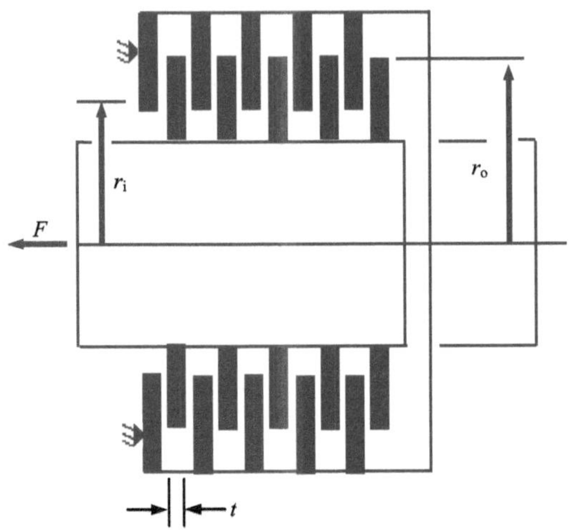

Fig. 5.9 Multiple disc clutch brake design.

Table 5.9 Comparisons of statistical results using reported optimizers in literature for multiple disc clutch brake design.

Methods	Worst	Mean	Best	Std	FEs
ABC (Rao et al., 2011)	0.352864	0.324751	0.313657	0.54	600
TLBO (Rao et al., 2011)	0.392071	0.3271662	0.313657	0.67	600
AEO	0.3332601	0.3216842	0.3136566	0.008154	500

Table 5.10 Comparisons of best solutions offered by reported optimizers for multiple disc clutch brake design.

	NSGA-II (Deb and Srinivasan, 2011)	TLBO (Rao et al., 2011)	ABC	PSO	CS	GSA	AEO
$x_1(r_i)$	70.0000	70	70	70	70	72	70
$x_2(r_o)$	90.0000	90	90	90	90	92	90
$x_3(t)$	1.50000	1	1	1	1	2	1
$x_4(F)$	1000.0000	810	790	860	810	815	810
$x_5(Z)$	3.0000	3	3	3	3	3	3
g_1	0.0000	0	0	0	0	0	0
g_2	22.0000	24.000000	24	24	24	22	24
g_3	0.9005	0.919427	0.9214172	0.9144542	0.9194278	0.9213930	0.9194278
g_4	9.7906	9830.371000	9.8345595	9.8199002	9.8303711	9.8304134	9.8303711
g_5	7.8947	7894.696500	7.8946966	7.8946966	7.8946966	7.8426028	7.8946966
g_6	3.3527	0.702013	0.3510866	1.5099273	0.7020132	1.0372174	0.7020132
g_7	60.6250	37,706.250000	35.2937500	43.7375000	37.7062500	40.1239024	37.7062500
g_8	11.6473	14.297986	14.6489134	13.4900727	14.2979868	13.9627826	14.2979868
f_5	0.4704	0.3136566	0.3136570	0.3136566	0.3136566	0.3175771	0.3136566

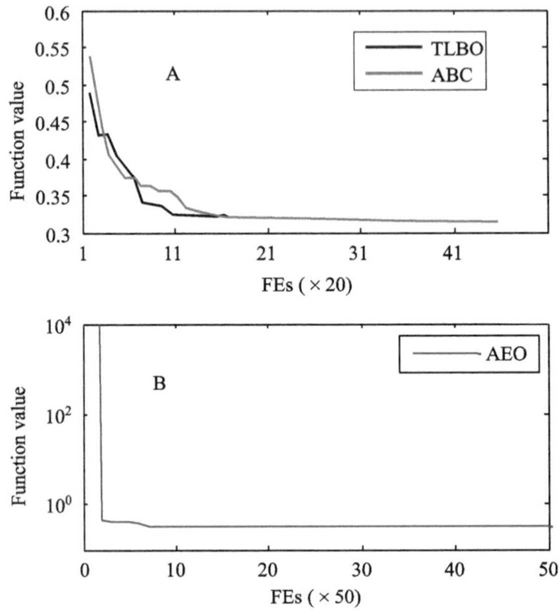

Fig. 5.10 Convergence curves of algorithms for multiple disc clutch brake design problem. (A) TLBO and ABC, (B) AEO.

power generation cost of the power system (Panigrahi et al., 2008; Hardiansyah, 2013; Orike, 2015; Nawaz et al., 2017; Mohit and Dhillon, 2018). The static economic load dispatch (ELD) problem concerns minimizing the fuel cost of generating units for a specific period of operation, which is essentially a multiconstrained, highly nonlinear nonconvex optimization problem. The objective function of ELD can be defined as (Das and Suganthan, 2012):

$$\text{Minimize } F = \sum_{i=1}^{N_G} f_i(P_i) \tag{5.1}$$

$$f_i(P_i) = a_i P_i^2 + b_i P_i + c_i, i = 1, 2, 3, \cdots, N_G \tag{5.2}$$

where a_i, b_i and c_i are the cost coefficients of ith generating unit. P_i is the real power output (in MW) of ith generator corresponding to time period t. N_G is the number of online generating units to be dispatched.

The cost function for unit with valve point loading effect is given as:

$$f_i(P_i) = a_i P_i^2 + b_i P_i + c_i + \left| e_i \sin\left[f_i\left(P_i^{\min} - P_i \right) \right] \right| \tag{5.3}$$

where e_i and f_i are the cost coefficients corresponding to valve point loading effect.

Power balance constraints or demand constraints are based on the principle of equilibrium between total system generation as follows:

$$\sum_{i=1}^{N_G} P_i = P_D + P_L \tag{5.4}$$

where P_D is the total system loads and P_L is the losses obtained using B-coefficients:

$$P_L = \sum_{i=1}^{N_G} \sum_{j=1}^{N_G} P_i B_{ij} P_j + \sum_{i=1}^{N_G} B_{0i} P_i + B_{00} \tag{5.5}$$

Generator constraints are represented by a pair of inequality constraints:

$$P_i^{\min} \leq P_i \leq P_i^{\max} \tag{5.6}$$

where $P_i^{\min}$and $P_i^{\max}$ are lower and upper bounds for power outputs of the ith generating unit.

The inclusion of ramp rate limits is modified to:

$$\max\left(P_i^{\min}, UR_i - P_i\right) \leq P_i \leq \max\left(P_i^{\max}, P_i^{t-1} - DR_i\right) \tag{5.7}$$

where P_i^{t-1} is the power generation of ith unit at previous hour and UR_i and DR_i are the upper and lower ramp rate limits respectively.

To avoid operation in these zones in order to economize the production, the following constraint is given:

$$P_i \leq p_{\text{lower}} \text{ and } P_i \geq p_{\text{upper}} \tag{5.8}$$

where p_{lower} and p_{upper} are the lower and upper limits of a given prohibited zone for ith generating unit.

5.2.2 Results and discussion

To verify the effectiveness of the AEO algorithm, the ELD problem is solved by the AEO algorithm and other algorithms such as PSO, GA, ABC, and their results were compared. The population size of 50 and maximum number of iterations of 1000 are considered. The ELD problem with different number of units, viz., 6, 13, 15, 40, and 140 is considered by these optimizers. Figs. 5.11–5.15 show the convergence curves of different algorithms for the ELD with different number of units. Tables 5.11–5.14 show the optimal power outputs and the costs corresponding to different number of units using different algorithms. From these figures and tables, the AEO algorithm

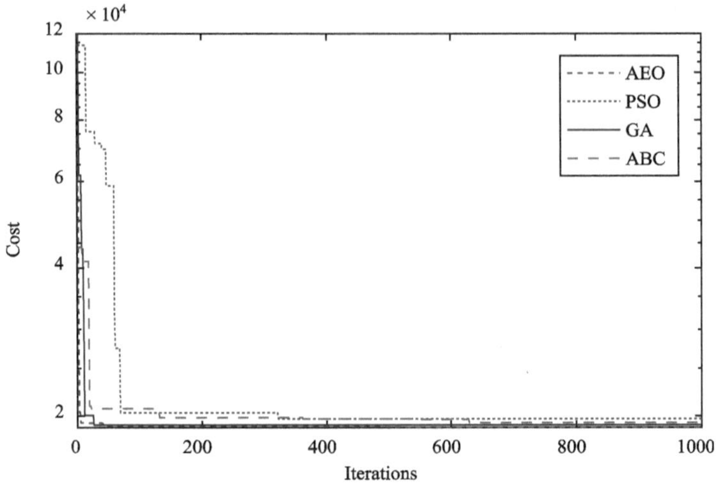

Fig. 5.11 Convergence curves of different algorithms for ELD with 6 units.

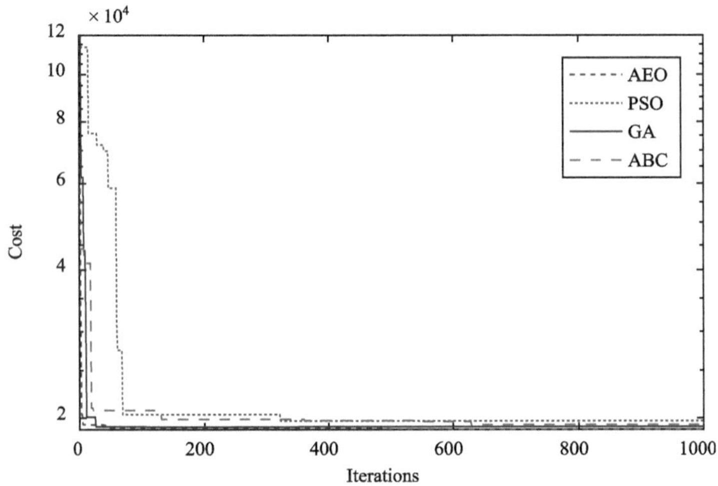

Fig. 5.12 Convergence curves of different algorithms for ELD with 13 units.

is more competitive to provide optimal power outputs for ELD with different number of units. Meanwhile, the AEO algorithm exhibits a better convergence rate compared to its competitors. Therefore, it can be concluded that the AEO algorithm is competitive and effective when solving the ELD problem.

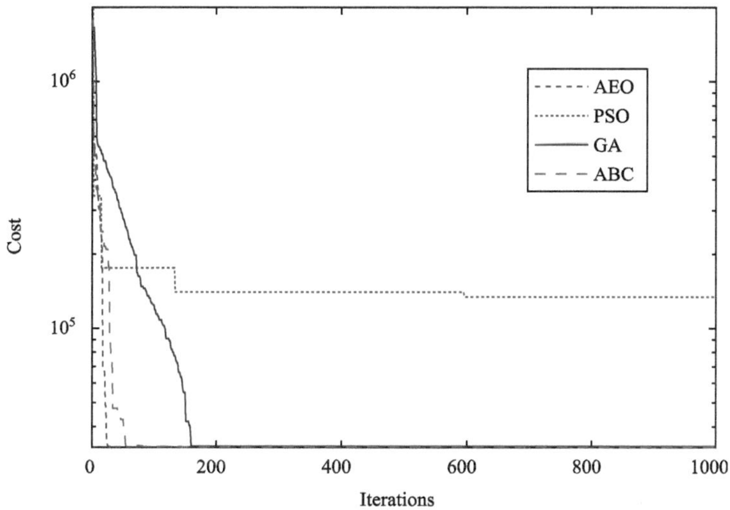

Fig. 5.13 Convergence curves of different algorithms for ELD with 15 units.

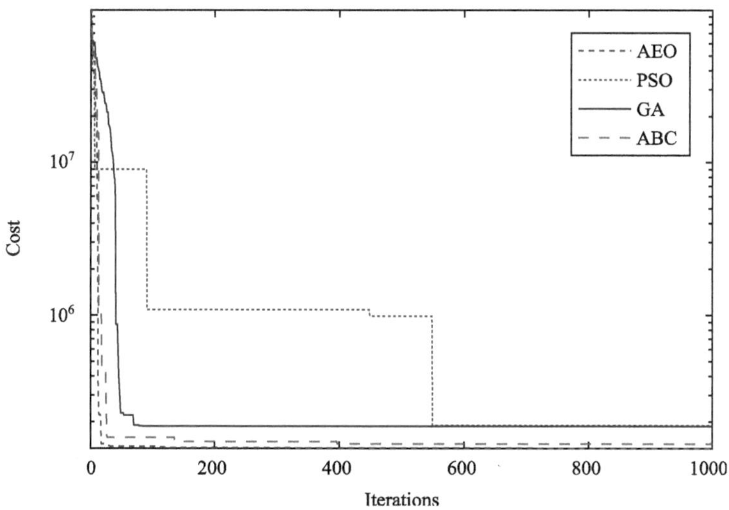

Fig. 5.14 Convergence curves of different algorithms for ELD with 40 units.

5.3 Hydrothermal scheduling problem

5.3.1 Problem description

Hydrothermal scheduling is a very complex problem which involves nonlinear relationships of the decision variables, the cascaded nature of a hydraulic network, water carry delays, and time links between consecutive

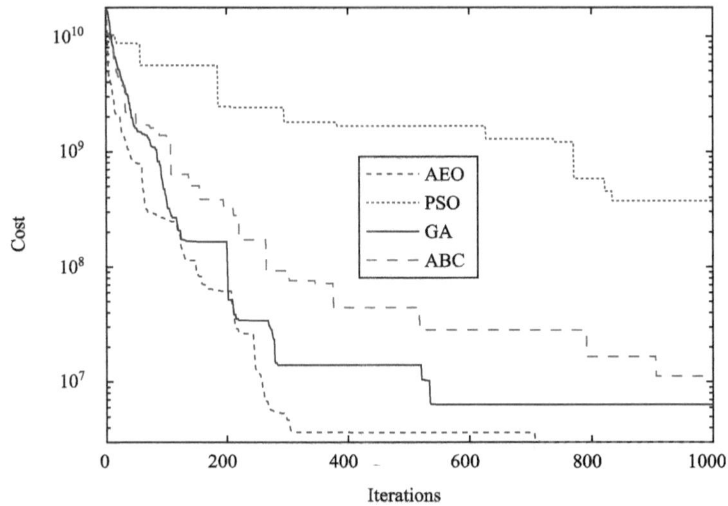

Fig. 5.15 Convergence curves of different algorithms for ELD with 140 units.

Table 5.11 Results comparisons of different algorithms for ELD with 6 units.

Algorithms	AEO	PSO	GA	ABC
P1	456.2434	462.5801	445.0516	447.6452
P2	171.1281	195.4907	183.6101	185.4603
P3	258.8859	201.7352	202.5682	253.7856
P4	149.7126	141.5372	138.7295	137.9717
P5	164.4873	176.1589	194.1577	160.4169
P6	74.8540	98.1313	111.9050	90.1952
Cost	15,446.4243	15,568.5381	15,495.6150	15,476.7460

Table 5.12 Results comparisons of different algorithms for ELD with 13 units.

Algorithms	AEO	PSO	GA	ABC
P1	182.1055	358.2542	630.7989	446.4334
P2	226.1063	151.9809	189.4189	98.2008
P3	287.5743	165.8800	152.6592	158.1206
P4	97.1810	96.2676	79.8605	152.1156
P5	156.1689	137.4162	132.3924	123.8360
P6	93.1575	95.1422	63.6617	115.1793
P7	136.1529	167.0966	74.7534	102.3941
P8	111.2353	164.8851	73.9652	146.9429
P9	144.5155	109.9555	92.5311	119.9092
P10	120.0000	108.4802	74.3038	86.6497
P11	63.4973	89.4710	90.4944	75.6865
P12	91.8915	73.0225	84.9805	74.3945
P13	90.4140	82.1412	60.1800	100.1342
Cost	18,943.1951	19,724.9863	19,187.1155	19,384.9633

Table 5.13 Results comparisons of different algorithms for ELD with 15 units.

Algorithms	AEO	PSO	GA	ABC
P1	384.3372	450.2498	379.3170	429.8935
P2	379.8906	293.1641	304.1729	365.5997
P3	128.8293	67.5165	128.7125	125.9981
P4	121.4230	92.9262	126.6648	115.4025
P5	169.1457	163.9962	167.9412	162.8212
P6	429.1974	429.2693	362.9116	405.6860
P7	426.1594	413.1373	395.6456	398.9194
P8	113.1822	104.9852	154.4304	143.1164
P9	128.4585	133.0867	159.8144	132.9983
P10	157.4256	141.6701	156.5392	146.9291
P11	79.9435	73.9933	76.0658	76.3904
P12	49.0440	51.6946	77.1537	70.2898
P13	25.0001	50.7729	80.9470	29.9463
P14	51.1458	48.9910	51.5093	22.8709
P15	24.1671	45.9064	50.4108	40.9612
Cost	32,971.9072	134,110.3516	33,323.1167	33,078.5004

Table 5.14 Results comparisons of different algorithms for ELD with 40 units.

Algorithms	AEO	PSO	GA	ABC
P1	77.9588	108.1752	77.6450	83.8140
P2	69.6665	80.5659	51.0228	81.3703
P3	81.4773	88.5691	94.2001	85.6019
P4	149.2282	183.7776	176.2197	125.3655
P5	61.5544	82.8755	91.4383	88.9795
P6	106.1301	117.4927	125.6383	113.7718
P7	258.0158	258.3579	254.2503	254.4473
P8	297.6924	253.5883	238.7256	279.1783
P9	284.6008	247.4730	279.9390	271.4820
P10	239.9601	224.3766	201.1547	253.0023
P11	315.6537	270.4223	205.2224	272.8996
P12	223.5494	329.8077	361.3298	136.1792
P13	408.8528	430.1867	460.7017	329.1383
P14	361.5744	440.3181	424.5773	453.4995
P15	469.7164	378.2147	433.5292	499.3484
P16	470.9030	331.6035	413.9017	479.1414
P17	494.0123	452.0224	427.6032	416.2076
P18	457.9657	462.7872	396.4959	405.8983
P19	515.4995	478.8891	421.5229	522.5991
P20	529.1886	541.3778	520.5268	538.8607
P21	434.1712	363.7145	525.8432	508.6085
P22	457.9973	512.3035	397.3104	474.7955

Continued

Table 5.14 Results comparisons of different algorithms for ELD with 40 units—cont'd

Algorithms	AEO	PSO	GA	ABC
P23	521.6238	294.9351	496.8430	474.3583
P24	424.3343	402.2520	352.5682	499.8455
P25	550.0000	534.7725	433.1607	402.7378
P26	483.3877	534.4148	443.3595	523.1715
P27	48.0468	129.1910	89.2527	97.6058
P28	42.3192	138.2502	114.5474	93.7261
P29	77.4338	128.6704	120.5593	79.8286
P30	72.8367	74.2071	112.3498	72.5794
P31	105.4596	126.4791	158.4038	108.6250
P32	135.1387	176.4575	195.2188	139.1279
P33	150.0257	178.6333	172.6096	154.5738
P34	160.8279	150.1137	148.5470	189.3145
P35	133.1634	167.0664	185.7397	181.1056
P36	170.1285	165.4764	185.6194	146.6717
P37	73.3886	56.3913	95.1408	62.3345
P38	95.2493	67.1427	54.0262	33.7522
P39	31.3868	53.2671	96.3173	78.8321
P40	459.8805	485.7164	466.9385	487.5959
Cost	133,983.6616	188,688.5828	186,990.2817	143,296.3109

schedules. In addition, considering the fact that we have to schedule the water discharges of 4 hydro units for 24 h, the dimension of the problem is 96 (Orero and Irving, 1998; Sivasubramani and Shanti, 2011; Das and Suganthan, 2012). Consequently, it is extremely different for many optimization algorithms to find a global optimal solution.

The total fuel cost for operation of the thermal system is considered as an objective function as follows (Das and Suganthan, 2012):

$$\text{Minimize } F = \sum_{i=1}^{M} f_i(P_{Ti}) \tag{5.9}$$

$$f_i(P_{Ti}) = a_i P_{Ti}^2 + b_i P_{Ti} + c_i + \left| e_i \sin\left[f_i\left(P_{Ti}^{\min} - P_{Ti} \right) \right] \right| \tag{5.10}$$

where f_i is the cost function corresponding to the equivalent thermal unit's power generation P_{Ti} at ith interval. M is the total number of intervals considered for the short-term schedule.

The demand constraint of the total power produced is represented as:

$$P_{T(i)} \sum_{k=1}^{N} P_{H(k,i)} = P_{D(i)} + P_{\text{Loss}(i)} \tag{5.11}$$

where $P_{H(k,i)}$ is the power generated by kth hydro unit at ith interval, $P_{D(i)}$ and $P_{Loss(i)}$ are the power demand and power loss at the ith interval, respectively, and N is the total number of hydro units.

The lower and upper limits are given for the thermal generator as:

$$P_T^{\min} \leq P_{T(i)} \leq P_T^{\max} \tag{5.12}$$

The lower and upper limits for the hydro plant's power generation are:

$$P_{H(k)}^{\min} \leq P_{H(k,\,i)} \leq P_{H(k)}^{\max} \tag{5.13}$$

The lowest and highest capacity limits of the volume of each reservoir's storages are represented as:

$$V_{(k)}^{\min} \leq V_{(k,\,i)} \leq V_{(k)}^{\max} \tag{5.14}$$

The restrictions of the reservoirs on the initial and final storage volume are:

$$V_{(k,\,0)} = V_{(k)}^{\text{initial}}, \ V_{(k,\,M)} = V_{(k)}^{\text{final}} \tag{5.15}$$

The lower and upper limits for water discharge rate of each reservoir are:

$$Q_{(k)}^{\min} \leq Q_{(k,\,i)} \leq Q_{(k)}^{\max} \tag{5.16}$$

The continuity equation for the volume stored in kth reservoir for the $(i+1)$th interval is:

$$V_{(k,\,i+1)} = V_{(k,\,i)} + \sum_{j=\Omega(k)} \left(Q_{(j,\,i-\tau)} + S_{(j,\,i-\tau)}\right) - Q_{(k,\,i)} - S_{(k,\,i)} + R_{(k,\,i)} \tag{5.17}$$

where $\Omega(k)$is the index set of the upstream reservoirs contributing to kth reservoir, τ is the time delay occurring for the water in jth upstream reservoir to reach kth reservoir. S and R are the spillage and inflow rates, respectively.

5.3.2 Results and discussion

In order to test the optimization performance of the AEO algorithm, three cases of hydrothermal scheduling problem were used. The results provided by PSO, GA, and ABC were compared. The number of population size is 50, and the maximum of iteration of the algorithms is 1000. The convergence curves obtained by different algorithms for three different cases are represented in Figs. 5.16–5.18 and the comparison of the total cost provided by the AEO algorithm and other algorithms for three different cases are

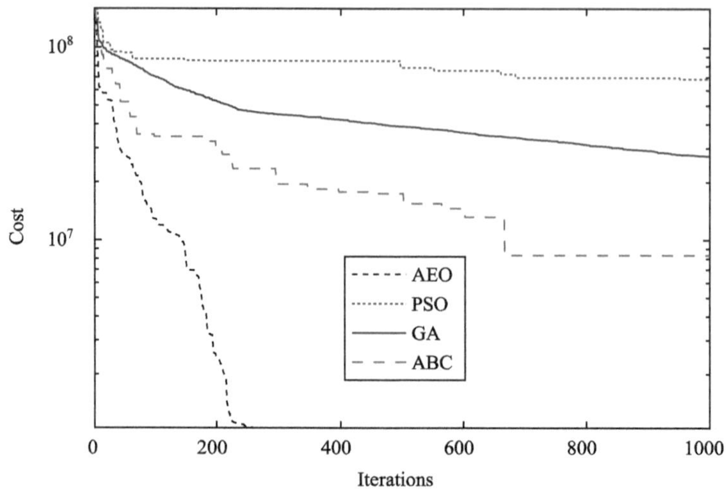

Fig. 5.16 Convergence curve different algorithms for Case 1.

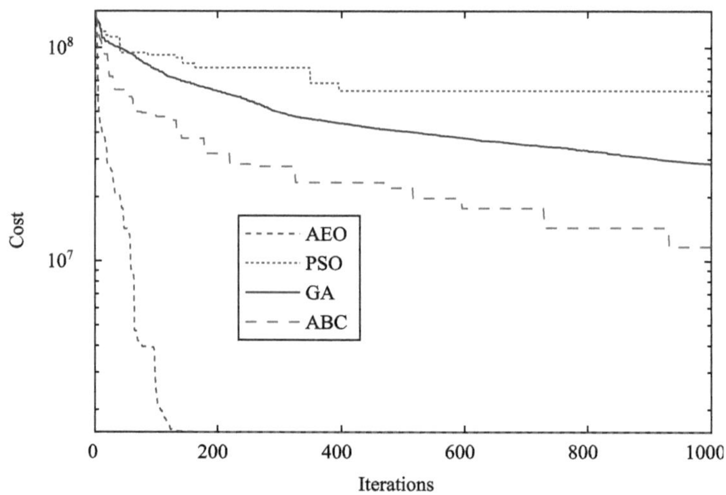

Fig. 5.17 Convergence curve different algorithms for Case 2.

shown in Table 5.15. The best hourly plant discharges provided by the AEO method are given in Tables 5.16–5.18. From these results, the AEO algorithm has searched better optimal solutions for the hydrothermal scheduling problem for different cases. Additionally, the AEO algorithm has a better convergence rate for three cases compared to other algorithms.

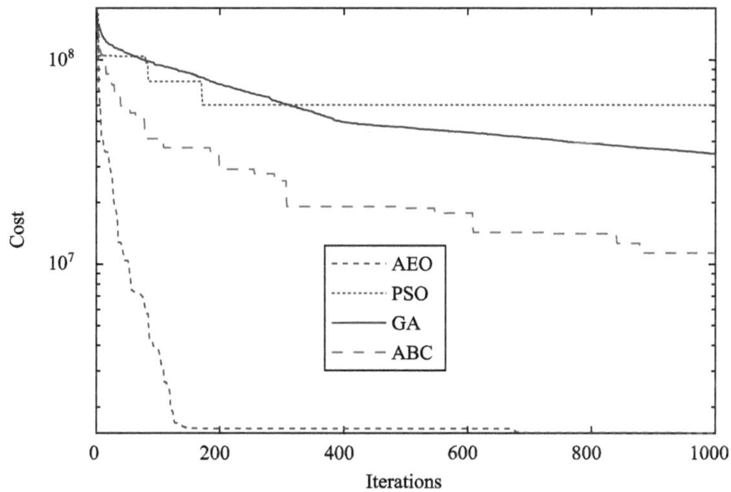

Fig. 5.18 Convergence curve different algorithms for Case 3.

Table 5.15 The optimal cost provided by different algorithms for different cases.

Algorithms	AEO	PSO	GA	ABC
Case 1	1.0536E+06	6.8658E+07	2.7368E+07	8.3272E+06
Case 2	1.5704E+06	6.2980E+07	2.8187E+07	1.1658E+07
Case 3	1.4901E+06	6.0319E+07	3.4763E+07	1.1334E+07

Table 5.16 Hourly plant discharges provided by AEO for Case 1.

Time/h	$Q_1/(10^4\ m^3/h)$	$Q_2/(10^4\ m^3/h)$	$Q_3/(10^4\ m^3/h)$	$Q_4/(10^4\ m^3/h)$
1	12.0866	6.4130	25.8945	13.0064
2	5.1289	6.4841	21.9617	17.6141
3	8.4805	6.4753	17.6761	13.6694
4	11.0014	6.0308	15.6707	13.0465
5	9.6030	8.6546	11.0307	13.0235
6	9.2765	7.8773	20.6541	13.7452
7	5.1465	8.0839	13.4322	13.5967
8	9.8728	9.4388	24.4601	13.3438
9	10.6226	8.3359	15.5310	13.4120
10	5.0883	12.4815	10.0831	15.0853
11	10.0811	10.5526	18.5152	14.9154
12	5.0393	10.0072	19.1753	16.2709

Continued

Table 5.16 Hourly plant discharges provided by AEO for Case 1—cont'd

Time/h	$Q_1/(10^4\ m^3/h)$	$Q_2/(10^4\ m^3/h)$	$Q_3/(10^4\ m^3/h)$	$Q_4/(10^4\ m^3/h)$
13	11.0266	9.5883	14.4471	14.0271
14	9.5844	7.1658	15.1218	14.4381
15	9.7958	8.4846	18.8580	13.5119
16	5.7279	7.7813	19.7138	17.1190
17	5.3187	6.2837	24.5215	14.8925
18	10.9040	6.2477	10.5650	15.8290
19	7.2354	8.5090	21.0627	14.8796
20	6.7986	12.4997	22.8110	13.5644
21	6.4126	12.4523	16.5696	14.3606
22	5.1521	6.0080	13.9834	14.9611
23	8.0326	7.9126	19.6919	13.7583
24	7.5838	8.2320	10.1005	15.9148

Table 5.17 Hourly plant discharges provided by AEO for Case 2.

Time/h	$Q_1/(10^4\ m^3/h)$	$Q_2/(10^4\ m^3/h)$	$Q_3/(10^4\ m^3/h)$	$Q_4/(10^4\ m^3/h)$
1	6.7504	6.7060	16.6107	14.5049
2	9.5887	9.0197	27.8033	15.2381
3	7.7933	9.4282	20.1028	13.0657
4	10.3123	9.6883	16.8671	14.2519
5	10.0844	12.3987	24.1003	15.9850
6	7.6972	8.2881	22.6087	14.9612
7	7.7026	8.4698	17.0810	16.6511
8	7.9577	8.8615	17.1312	14.1678
9	7.0657	7.1195	19.7130	16.4924
10	7.5682	10.2431	16.4424	16.7338
11	9.7882	6.9078	17.1319	13.3515
12	7.3539	9.4716	17.1767	13.5729
13	9.6617	6.2667	18.7699	14.0419
14	9.2304	7.4823	14.5301	16.1045
15	6.0254	8.1578	20.6401	18.0199
16	8.8982	6.9922	19.5896	13.3256
17	7.5989	6.6526	19.2767	13.4714
18	11.1632	6.9504	15.5918	13.0441
19	8.1381	10.3677	13.8391	14.9375
20	6.6808	8.3019	12.2201	15.2919
21	7.2622	8.1067	10.0133	13.3449
22	5.0243	8.5158	13.8121	14.5243
23	8.2244	9.1789	10.2852	13.5207
24	7.4298	8.4247	16.1893	15.4235

Table 5.18 Hourly plant discharges provided by AEO for Case 3.

Time/h	$Q_1/(10^4\ m^3/h)$	$Q_2/(10^4\ m^3/h)$	$Q_3/(10^4\ m^3/h)$	$Q_4/(10^4\ m^3/h)$
1	6.5690	7.1815	25.3931	13.8428
2	6.1984	6.9459	20.0623	13.2362
3	10.5303	7.5969	15.0884	14.1338
4	6.4486	6.5928	16.2236	13.1145
5	7.6454	10.1015	15.2198	14.4635
6	10.2359	8.7204	19.7079	15.4010
7	7.1715	7.1148	18.9431	15.3292
8	6.3964	7.8714	18.2762	13.1687
9	5.8569	8.9176	15.9773	13.0840
10	7.6096	8.5638	13.5131	13.1163
11	10.4897	6.6109	10.3868	14.9727
12	9.2560	12.4152	18.2669	15.9463
13	6.9473	10.6409	12.1672	14.4835
14	9.1739	6.5095	23.1770	16.2120
15	7.8352	8.6709	21.1080	14.0197
16	8.4474	8.5688	19.2927	21.1902
17	10.1351	6.8010	17.8590	13.6645
18	10.4879	7.6041	20.0667	13.3633
19	8.0207	10.8205	21.1245	13.8856
20	8.2820	8.6920	16.3900	13.0868
21	6.4831	7.3859	16.7219	13.0188
22	7.1681	7.0656	11.5855	14.0409
23	6.8592	11.7290	10.8455	13.6829
24	10.7517	8.8804	16.6182	14.5870

5.4 Conclusions

Based on five challenging engineering design problems, the statistical results with different indices suggest that, compared with other reported popular methods, AEO is more promising for real-world problems requiring a faster convergence rate and less computational expense with a specified precision for final solutions. The applications of AEO to the hydrothermal scheduling problem and static economic load dispatch problem were also investigated in this study. It is difficult for some optimization techniques to solve these problems because of their particularity and unknown search space. However, AEO can find better optimal solutions effectively, proving that the proposed algorithm is highly promising in different engineering fields.

Generally speaking, AEO provides enough competitive results compared to other considered metaheuristic methods on the reported results in the literature and the experimental results in this study. However, the computational efficiency and solution quality depend largely on the presentation and complexity of underlying problems. This is a general characteristic not just for AEO, but for most metaheuristics. Therefore, our algorithm tends to handle the real-world problems focusing on convergence efficiency and computational expense with an acceptable precision for final solutions.

References

Akay, B., Karaboga, D., 2012. Artificial bee colony algorithm for large-scale problems and engineering design optimization. J. Intell. Manuf. 23 (4), 1001–1014.

Askarzadeh, A., 2016. A novel metaheuristic method for solving constrained engineering optimization problems: crow search algorithm. Comput. Struct. 169, 1–12.

Belegundu, A.D., 1982. A Study of Mathematical Programming Methods for Structural Optimization. Department of Civil and Environmental Engineering, University of Iowa, Iowa City, IA.

Coello, C.A.C., 2000. Use of a self-adaptive penalty approach for engineering optimization problems. Comput. Ind. 41 (2), 113–127.

Coello, C.A.C., Becerra, R.L., 2004. Efficient evolutionary optimization through the use of a cultural algorithm. Eng. Optim. 36, 219–236.

Coello, C.A.C., Montes, E.M., 2002. Constraint-handling in genetic algorithms through the use of dominance-based tournament selection. Adv. Eng. Inform. 16, 193–203.

Das, S., Suganthan, P.N., 2012. Problem Definitions and Evaluation Criteria for CEC 2011 Competition on Testing Evolutionary Algorithms on Real World Optimization Problems, Technical Report, 2012. pp. 1–42.

Dos Santos Coelho, L., 2010. Gaussian quantum-behaved particle swarm optimization approaches for constrained engineering design problems. Expert Syst. Appl. 37 (2), 1676–1683.

Hardiansyah, H., 2013. Artificial bee colony algorithm for economic load dispatch problem. IAES Int. J. Artif. Intell. 2 (2), 90–98.

He, Q., Wang, L., 2006. An effective co-evolutionary particle swarm optimization for engineering optimization problems. Eng. Appl. Artif. Intell. 20, 89–99.

He, Q., Wang, L., 2007. A hybrid particle swarm optimization with a feasibility-based rule for constrained optimization. Appl. Math. Comput. 186, 1407–1722.

Huang, F.Z., Wang, L., He, Q., et al., 2007. An effective co-evolutionary differential evolution for constrained optimization. Appl. Math. Comput. 186 (1), 340–356.

Kannan, B.K., Kramer, S.N., 1994. An augmented Lagrange multiplier based method for mixed integer discrete continuous optimization and its applications to mechanical design. J. Mech. Des. 116, 405–411.

Mezura-Montes, E., Coello, C.A.C., 2005. Useful infeasible solutions in engineering optimization with evolutionary algorithms, MICAI 2005. Lect. Notes Artif. Int. 3789, 652–662.

Mirjalili, S., Lewis, A., 2016. The whale optimization algorithm. Adv. Eng. Softw. 95, 51–67.

Mohit, K., Dhillon, J.S., 2018. Hybrid artificial algae algorithm for economic load dispatch. Appl. Soft Comput. 71, 89–109.

Montes, E., Coello, C.A.C., Reyes, J.V., et al., 2006. Increasing successful offspring and diversity in differential evolution for engineering design. In: Proceedings of the Seventh

International Conference on Adaptive Computing in Design and Manufacture, pp. 131–139.
Nawaz, A., Saleem, N., Mustafa, E., Khan, U.A., 2017. An efficient global technique for solving network constrained static and dynamic economic dispatch problem. Turk. J. Electr. Eng. Comput. Sci. 25 (1), 73–82.
Ngo, T.T., Sadollah, A., Kim, J.H., et al., 2016. A cooperative particle swarm optimizer with stochastic movements for computationally expensive numerical optimization problems. J. Comput. Sci. 13, 68–82.
Orero, S.O., Irving, M.R., 1998. A genetic algorithm modeling framework and solution technique for short term optimal hydrothermal scheduling. IEEE Trans. Power Syst. 13, 501–518. https://doi.org/10.1109/59.667375.
Orike, S., 2015. Smart evolutionary algorithm for static economic load dispatch optimization in a thermal generating station. J. Am. Stat. Assoc. 146 (471), 7–22.
Osyczka, A., 2002. Evolutionary Algorithms for Single and Multicriteria Design Optimization: Studies in Fuzzyness and Soft Computing. PhysicaVerlag, Heidelberg.
Panigrahi, B.K., Pandi, V.R., Das, S., 2008. Adaptive particle swarm optimization approach for static and dynamic economic load dispatch. Energy Convers. Manag. 49 (6), 1407–1415.
Parsopoulos, K.E., Vrahatis, M.N., 2005. Unified particle swarm optimization for solving constrained engineering optimization problems. In: International Conference on Natural Computation. Springer, Berlin, Heidelberg, pp. 582–591.
Rao, R.V., Savsani, V.J., Vakharia, D.P., 2011. Teaching-learning-based optimization: a novel method for constrained mechanical design optimization problems. Comput. Aided Des. 43 (3), 303–315.
Ray, T., Liew, K.M., 2003. Society and civilization:an optimization algorithm based on the simulation of social behavior. IEEE Trans. Evol. Comput. 7 (4), 386–396.
Sivasubramani, S., Shanti, S.S., 2011. Hybrid DE–SQP algorithm for non-convex short term hydrothermal scheduling problem. Energy Convers. Manag. 52, 757–761.
Wang, L., Li, L.P., 2010. An effective differential evolution with level comparison for constrained engineering design. Struct. Multidiscip. Optim. 41, 947–963.
Zhao, W., Wang, L., Zhang, Z., 2019. Artificial ecosystem-based optimization: a novel nature-inspired meta-heuristic algorithm. Neural Comput. & Applic. https://doi.org/10.1007/s00521-019-04452-x.

CHAPTER 6

Supply-demand-based optimization

Contents

6.1 Introduction

The term "optimization" refers to finding optimal solutions with the most cost-effective form in a given solution space under the given constraints, by either maximizing or minimizing its objective function. Over the past few decades, the substantial increase of optimization problems from different fields has made optimization techniques into a major research area for the scientific community. The conventional optimization techniques, including steepest descent method (Goldstein, 1965) and Newton's method (Weerakoon and Fernando, 2000), were developed. However, owing to the inherent requirement of derivative and local search stagnation, they tend to become less powerful when solving complex nonlinear functions and constraints with multiple peaks (Wang et al., 2018; Talbi, 2009). Consequently, some stochastic optimization methods have emerged and have been widely applied in a wide range of fields (Maddala and Rengaswamy, 2012; Ibrahima et al., 2014; Chana, 2013; Lai et al., 2016; Hung and Wu, 2017). Stochastic optimization methods use random variables to

New Optimization Algorithms and their Applications
https://doi.org/10.1016/B978-0-323-90941-9.00006-5

generate random objective functions and constraint values. Their outstanding merit is randomness, which is able to help these methods easily bypass local optima and explore the entire variable space. Another merit of stochastic optimization method is that they treat considered problems as a black box, which means that one only needs to lay emphasis on the input and output of objective functions instead of their gradient information. Additionally, stochastic optimization methods are highly versatile and flexible, implying their extendibility and practicability to different types of optimization problems.

General speaking, there are two different classifications in stochastic optimization methods. One classification is based on the number of their random solutions over the course of iterations; they are divided into single-solution-based and swarm-based algorithms. For the former, stochastic algorithms randomly generate a single solution and update it by introducing minor changes in an iterative process. Simulate annealing (SA) (Kirkpatrick et al., 1983) belongs to this type of method. For the latter, stochastic algorithms randomly initialize a set of solutions rather than one solution and then manipulate them at each stage. The major difference between single-solution-based and swarm-based algorithms is that the latter is able to share the solution information among the swarm and improve the quality of solutions by introducing major change in subsequent iterations. This assists swarm-based algorithms explore the search space and avoid the local optima. To solve various optimization problems, many different optimization techniques are developed, which motivated from all types of inspirations from nature.

Inspired by the supply-demand mechanism in economics, this study proposes a new metaheuristic algorithm called supply-demand-based optimization (SDO) (Zhao et al., 2019). SDO mathematically simulates the demand relation of consumers and the supply relation of producers.

6.2 Supply-demand-based optimization

6.2.1 Inspiration

According to economic theory, a commodity price and commodity quantity in a market might firstly be subject to periodic fluctuations and then gradually become stable at their respective equilibrium points (Ezekiel, 1938). This process generally depends on both the supply relation of producers and demand relation of consumers.

The supply-demand mechanism is an economic theory of price determination in a market economy. According to this theory, a commodity quantity in the next time q_{t+1} depending on its current price p_t in a market is determined by the supply relation of producers, that is $q_{t+1} = f(p_t)$, here f is a linear supply function. When the current commodity price increases in a market, the commodity quantity will increase in supply at the next time, so f is an increasing function. The commodity price p_{t+1} in the next time, depending on its quantity q_{t+1} in the same time, is determined by the demand relation of consumers, this is $p_{t+1} = g(q_{t+1})$, here g is a linear demand function. When the commodity quantity increases, its price in a market will decrease, so g is a decreasing function. With the oscillations, these two functions finally intersect at a point $P(x_0, y_0)$, which is called the equilibrium point. x_0 and y_0 are the equilibrium price and equilibrium quantity of a commodity, respectively.

The supply function f can be expressed as (Waugh, 1964):

$$q_{t+1} - q_0 = a(p_t - p_0) \tag{6.1}$$

The demand function g can be expressed as (Waugh, 1964):

$$p_{t+1} - p_0 = -b(q_{t+1} - q_0) \tag{6.2}$$

where t is the time, a and b are the linear coefficients.

Generally, the supply-demand mechanism has two modes. One is the stability mode as shown in Fig. 6.1A. When $|ab| < 1$, the supply function f is steeper than the demand function g, then the oscillations reduce in magnitude with each time, so the curve of the commodity price and quantity with respects to time tends to spiral inwards. After a certain time of iterations, the price and quantity converge to the equilibrium point (x_0, y_0). The other is the instability mode as shown in Fig. 6.1B. When $|ab| > 1$, the demand function g is steeper than the supply function f, then the oscillations increase in magnitude with each time, a curve of the price and quantity with respects to time tends to spiral onwards. Therefore, the price and quantity diverge from the equilibrium point (x_0, y_0) more and more as time passes. Because these spiral curves often look like a cobweb, this demand and supply mechanism is named the cobweb model in economics (Ezekiel, 1938). The cobweb model is a very famous economic theory in which the price fluctuation results in fluctuation in supply, thus causing a periodic rise or fall in price. Therefore, it is extensively used for studying the price fluctuation of various products in certain types of markets.

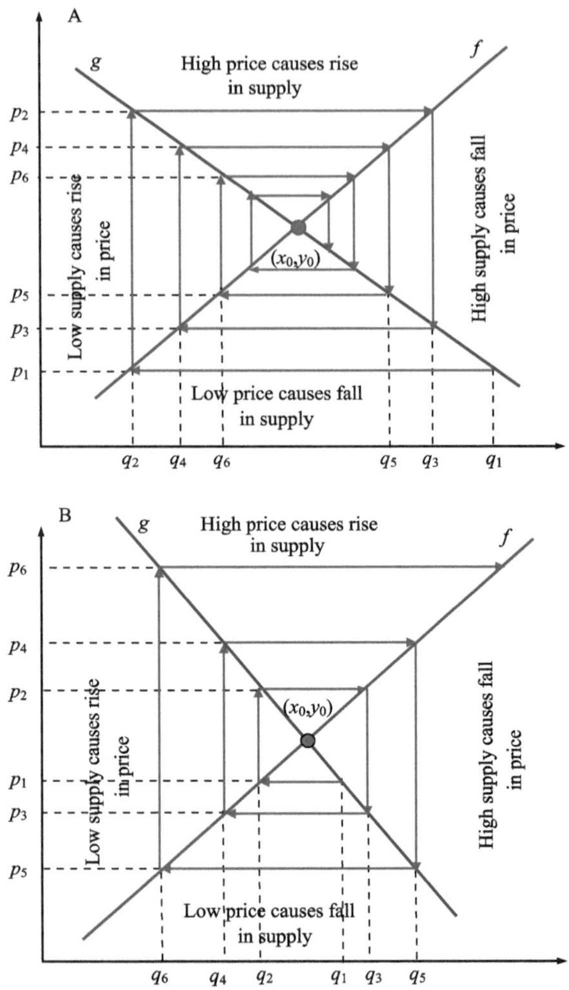

Fig. 6.1 Two modes of supply-demand mechanism: (A) stability mode, (B) instability mode.

6.2.2 Mathematical representation of SDO

According to the supply-demand mechanism, the stability mode can encourage both the commodity price and quantity to exploit the neighborhood of the equilibrium point and this exploiting process is oscillating in magnitude over time. The instability mode tends to force both the commodity price and quantity to gradually explore new regions far away from the equilibrium point. The stability mode gradually decreases oscillations in the supply-demand mechanism annd can easily be introduced to SDO as exploitation to perform a local search in a promising region. Similarly, the instability mode can be used for SDO as exploration to globally perform a search in the search space (Zhao et al., 2019). Fig. 6.2 shows the introduction from the supply-demand mechanism in economic theory to the proposed SDO algorithm.

For the SDO algorithm, suppose there are n markets, each of which has d types of different commodities, and each kind of commodity has a certain quantity and price. The d commodity prices of a market represent a candidate solution as d variables to the optimization problem, and the d commodity quantities of a market are reviewed as a possible candidate solution. If this possible candidate solution is better than the candidate solution, the candidate solution is replaced with the possible one. Because the proposed method is a swarm-based algorithm, the commodity price and the commodity quantity are given in two matrices, respectively.

The commodity price matrix of markets is given as:

$$X = \begin{bmatrix} x_1 \\ x_2 \\ \vdots \\ x_n \end{bmatrix} = \begin{bmatrix} x_1^1 \, x_1^2 \cdots x_1^d \\ x_2^1 \, x_2^2 \cdots x_2^d \\ \vdots \quad \vdots \quad \vdots \quad \vdots \\ x_n^1 \, x_n^2 \cdots x_n^d \end{bmatrix} \tag{6.3}$$

where d is the number of commodity prices (variables) in each market, n is the number of markets (candidate solutions), x_i^j $(i = 1,\cdots, n; j = 1,\cdots, d)$ is the jth commodity price in the ith market, and $\boldsymbol{x}_i$ $(i = 1,\cdots, n)$ is the ith commodity price vector corresponding to a candidate solution.

The commodity quantity matrix of markets is given as:

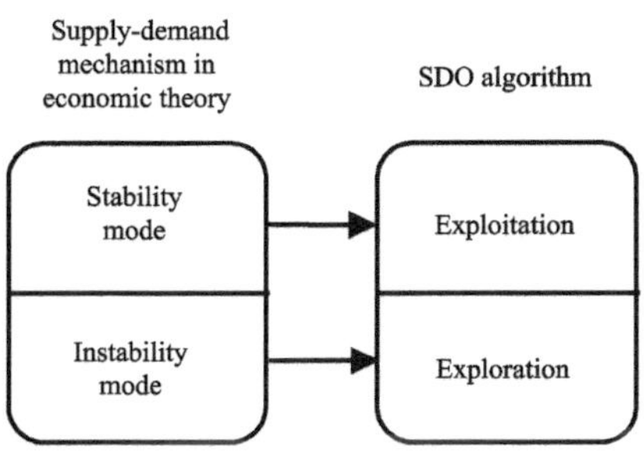

Fig. 6.2 Introduction form supply-demand mechanism to SDO.

$$Y = \begin{bmatrix} y_1 \\ y_2 \\ \vdots \\ y_n \end{bmatrix} = \begin{bmatrix} y_1^1 \, y_1^2 \cdots y_1^d \\ y_2^1 \, x_2^2 \cdots y_2^d \\ \vdots \;\; \vdots \quad \vdots \;\; \vdots \\ y_n^1 \, y_n^2 \cdots y_n^d \end{bmatrix} \tag{6.4}$$

where d is the number of commodity quantities in each market, y_i^j ($i = 1,\cdots, n$; $j = 1,\cdots, d$) indicates the jth commodity quantity in the ith market, and $\boldsymbol{y}_i$ ($i = 1,\cdots, n$) is the ith commodity quantity vector corresponding to a possible candidate solution.

The commodity price vector and quantity vector of each market are evaluated using the fitness function, respectively. The fitness value calculated by each price vector is stored as the return value of the fitness function for each market. Yet the fitness value calculated by each commodity quantity vector is also stored and only used for the solution replacement of a market according to its quality. For all the markets, an array for storing the fitness values of commodity price vectors in n markets is given:

$$Fx = [Fx_1 \; Fx_2 \cdots Fx_n]^T \tag{6.5}$$

Another array for storing the function values of commodity quantity vectors in n markets is:

$$Fy = [Fy_1 \; Fy_2 \cdots Fy_n]^T \tag{6.6}$$

where T signifies the transpose of the array.

Although the commodity price vector and quantity vector in each market are viewed as both solutions, the ways that they are treated and updated are different during the iterations. Meanwhile, the concepts of both the stability and instability modes in the supply-demand theory are utilized to perform exploration and exploitation in SDO, respectively.

The commodity equilibrium price $\boldsymbol{x}_0$ and equilibrium quantity $\boldsymbol{y}_0$ are two important components in SDO, yet they are not known a priori during the iterations, therefore, they need to be first designated. From the above section, it should be noted that the updates of both the demand function and supply function of a commodity is respectively based on both a fixed equilibrium price and a fixed equilibrium quantity from beginning to end. However, this updating will make the algorithm trap into the local extrema quickly and result in premature convergence. To prevent this, the equilibrium price of each commodity should be variable at each iteration, likewise with the equilibrium quantity of each commodity. Specifically, for each iteration, each market chooses a commodity quantity vector from the quantity array as its quantity equilibrium vector by means of its probability, and the better the fitness value of the quantity vector in a market, the higher the probability of that the quantity vector chosen is. Meanwhile, each market also chooses either a price vector from the price array according to its probability or the average of commodity price vectors of all the markets as the equilibrium price vector to improve exploration. The equilibrium quantity vector $\boldsymbol{y}_0$ is represented as follows:

$$N_i = \left| F\boldsymbol{y}_i - \frac{1}{n}\sum_{i=1}^{n} F\boldsymbol{y}_i \right| \tag{6.7}$$

$$Q = \frac{N}{\sum_{i=1}^{n} N_i} \tag{6.8}$$

$$y_0 = y_k, \; k = \text{Roulette wheel selection}(Q) \tag{6.9}$$

The equilibrium price vector $\boldsymbol{x}_0$ is represented as:

$$M_i = \left| F\boldsymbol{x}_i - \frac{1}{n}\sum_{i=1}^{n} F\boldsymbol{x}_i \right| \tag{6.10}$$

$$P = \frac{M}{\sum_{i=1}^{n} M_i} \tag{6.11}$$

$$\boldsymbol{x}_0 = \begin{cases} r_1 \cdot \dfrac{\sum_1^n \boldsymbol{x}_i}{n} & \text{if rand} < 0.5 \\ x_k, k = \text{Roulette wheel selection}(P) & \text{if rand} \geq 0.5 \end{cases} \tag{6.12}$$

where r_1 is a random number in [0, 1]. According to Eq. (6.12), SDO assumes that there is a probability of 50% to choose either the average of commodity price vector or a price vector depending on its probability in the price array as the equilibrium price vector. With the definitions of both the equilibrium price vector $\boldsymbol{x}_0$ and equilibrium quantity vector $\boldsymbol{y}_0$, the supply function and demand function are proposed as follows, respectively:

$$\boldsymbol{y}_i(t+1) = \boldsymbol{y}_0 + \alpha \cdot [\boldsymbol{x}_i(t) - \boldsymbol{x}_0] \tag{6.13}$$

$$\boldsymbol{x}_i(t+1) = \boldsymbol{x}_0 - \beta \cdot [\boldsymbol{y}_i(t+1) - \boldsymbol{y}_0] \tag{6.14}$$

where $\boldsymbol{x}_i(t)$ is the ith commodity price vector at the time t, $\boldsymbol{y}_i(t)$ is the ith commodity quantity vector at the time t, and α and β are the supply weight and demand weight, respectively.

From Eq. (6.13), the commodity quantity vector in each market can be updated according to both the equilibrium quantity vector and equilibrium price vector. Plugging Eqs. (6.13) into (6.14) the demand equation can be rewritten as:

$$\boldsymbol{x}_i(t+1) = \boldsymbol{x}_0 - \alpha\beta \cdot [\boldsymbol{x}_i(t) - \boldsymbol{x}_0] \tag{6.15}$$

It can be found that from this equation, the commodity price vector is updated with respect to its current price vector according to the equilibrium price vector actually. Different commodity price vectors can be obtained by adjusting the values of weights α and β. To perform exploration and exploitation in the SDO, both the supply weight α and demand weight β need to be appropriately presented. These two weights can be formulated as:

$$\alpha = \frac{2 \cdot (T - t + 1)}{T} \cdot \sin(2\pi r) \tag{6.16}$$

$$\beta = 2 \cdot \cos(2\pi r) \tag{6.17}$$

where T is the maximum number of iterations and r is a random number or vector in [0, 1]. Let the variable L equal the product of weights α and β, we can get:

$$L = \alpha\beta = \frac{4 \cdot (T - t + 1)}{T} \cdot \sin(2\pi r)\cos(2\pi r) \tag{6.18}$$

where $|L| < 1$ corresponds to the stability mode as depicted in Fig. 6.3A, different commodity price vectors around the equilibrium price $\boldsymbol{x}_0$ can be obtained with respect to the current price vector by adjusting the weights

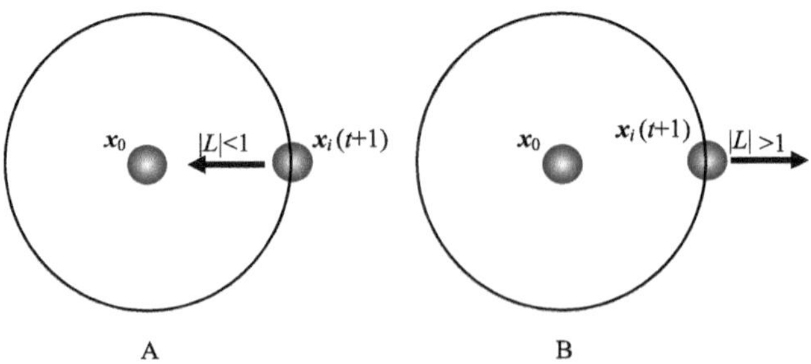

Fig. 6.3 Stability mode and instability mode of SDO algorithm: (A) stability mode, (B) instability mode.

α and β, and these price vectors are possible to be randomly changed between the current price vector and the equilibrium price vector by the random number or vector r. This is to say, the algorithm allows each market to update its all commodity prices in the neighborhood of the current commodity prices and mimic the stability mode in Fig. 6.1A. This mechanism emphasizes exploitation and encourages SDO algorithm to search locally. As depicted in Fig. 6.3B, $|L| > 1$ corresponds to the instability mode that allows the commodity price vector in any market to move far away from the equilibrium price vector, this forces each market to search new promising regions in the search space and simulate the instability mode in Fig. 6.1B. This mechanism concentrates on exploration and forces SDO algorithm to search globally.

The values of the variable L over iterations are depicted in Fig. 6.4 where the maximum number of iterations T is set as 1000. As observed in this figure, in early iterations, the values of L are greater than 1 or less than -1 with the high probability. As the iterations increase, this high probability begins to decrease, and there is an increasing probability that the function values of L are in $[-1, 1]$. In later iterations, the values of L are in $[-1, 1]$ with the increasingly high probability. Obviously, the SDO gets high exploration in early stage of iterations and smoothly switches to high exploitation in later stage of iterations.

After updating both the commodity price and commodity quantity vectors in each iteration, they need to be evaluated by the fitness function and the results are stored in their respective arrays. If the fitness value of the ith commodity quantity vector is better than that of the ith commodity price vector, the ith commodity price vector will be replaced with this quantity vector as a candidate solution. Fig. 6.5 shows the solution replacement mechanism.

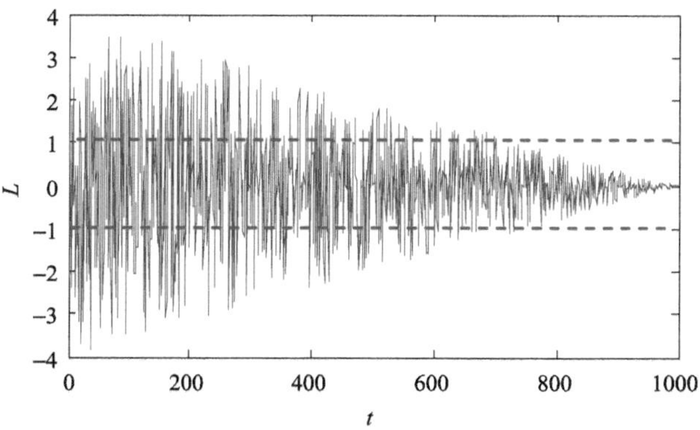

Fig. 6.4 Values of variable L over iterations.

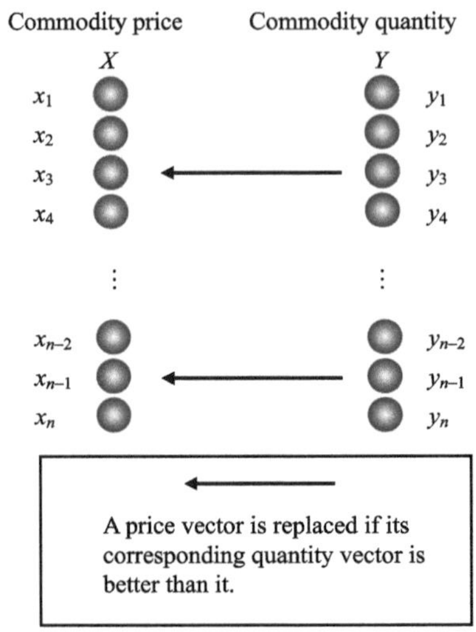

Fig. 6.5 Solution replacement mechanism.

SDO starts the optimization by creating a set of markets randomly. At each iteration, each commodity quantity of a market is updated with respect to both the equilibrium price and equilibrium quantity, and then each commodity price of a market is updated with respect to the equilibrium price. The equilibrium price vector can be randomly switched between a chosen commodity price vector form the price array according to its probability in the price array and the average of commodity price vectors. The equilibrium quantity vector is chosen from the quantity array according to its probability. These updates for commodity prices and quantities are achieved by adjusting the values of weights both α and β. The values of the variable L linearly decrease with random fluctuation in order to perform either exploration or exploitation. The commodity prices of a market are inclined to diverge from the equilibrium price when $|L| > 1$ and converge towards the equilibrium price when $|L| < 1$. Then the updated price vectors and quantity vectors are evaluated by the fitness function. For each market, if the fitness value of its commodity quantities is better than that of its commodity prices, its commodity prices will be replaced with its commodity quantities as a candidate solution. Eventually, the best commodity price vector of a market as the best solution found so far is returned when the stop criterion is satisfied. Fig. 6.6 gives the pseudo code of SDO algorithm and Fig. 6.7 shows an effective swarm search of SDO for both the unimodal and multimodal functions in a two-dimensional space.

```
Initialize the market population and weights: the
commodity price vectors xi and commodity quantity
vectors yi are randomly initialized, calculate their fitness
values Fxi and Fyi, replace xi by yi if Fyi is better Fxi, and
xbest= the best solution found so far. The weights α and β
are set.
  While the stop criterion is not satisfied do
    For each market (i=1,···,n)
      Determine the equilibrium quantity y0 by Equations
      (6.7–6.9) and the equilibrium price x0 by Equations
      (6.10–6.12), respectively.
      Update the commodity quantity vector yi by
      Equation (6.13).
      Update the commodity price vector xi by Equation
      (6.14).
      Calculate their fitness values Fxi and Fyi.
      If Fyi is better than Fxi,
         replace xi by yi,
      End If.
    End For.
    Update the best solution found so far xbest.
  End While.
Return the best solution found so far xbest.
```

Fig. 6.6 Pseudo code of SDO algorithm.

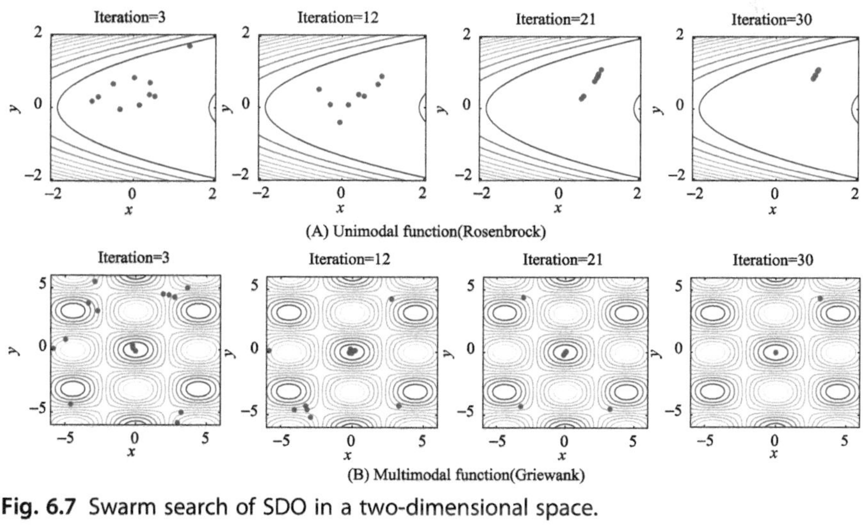

Fig. 6.7 Swarm search of SDO in a two-dimensional space.

6.3 Experimental results and discussion

6.3.1 Experimental setup

A set of comprehensive benchmark functions with known solutions are used to test the performance of algorithms (Zhao et al., 2019). We adopt 29 mathematical functions in literature (Zhao and Wang, 2016; Digalakis and Margaritis, 2011; Liang et al., 2014) for comparison. These functions are categorized into four types: unimodal, multimodal, fixed-dimensional, and composite. The unimodal functions (f_1–f_7) in Table A.1 have only one global optimum and no local optima, hence the convergence rate and exploitation of algorithms can be examined. Yet multimodal functions (f_8–f_{13}) in Table A.2 have a considerable number of local optima and low-dimensional functions (f_{14}–f_{23}) in Table A.3 have few local optima, so both are used to benchmark exploration, and local extrema avoidance of algorithms. Finally, the composite functions (f_{24}–f_{29}) in Table A.6 are complex multimodal functions combining shifted, rotated, expanded, and biased functions, and their detailed description in the CEC 2005 special session is available in Nowcki (1974). These composite functions are more challenging than unimodal and multimodal functions because they can shift the global optima from the specific position to random position before each iteration and occasionally relocate the global optima on the boundary of search space. Therefore, they are especially adapted to test the balance between exploration and exploitation of algorithms.

In order to test the performance of the proposed SDO algorithm, some classic and recently proposed stochastic algorithms are employed for comparison: GA, PSO, DE, CS, ABC, and GSA. Although numerous variants of algorithms have been developed, the comparisons of standard versions can be used to interpret the results of larger groups. In this test, the population size is set as 50 for all the tested algorithms. In addition, every algorithm runs 30 times for each function and implements 50,000 function evaluations (FEs) in each run. The results are based on the average performance of these runs. Two performance evaluation indexes are used to quantitatively compare all the algorithms: The average and standard deviation of best-so-far solutions. The parameter values of each algorithm are presented as follows:

(1) PSO: Inertia coefficient linearly reduces from 0.9 to 0.2, acceleration coefficients $c_1 = 2$ and $c_2 = 2$;

(2) GA: Decreasing coefficient $\gamma = 20$, mutation rate $p_m = 0.2$, crossover rate $p_c = 0.8$ and crossover adopts roulette wheel method;

(3) DE: Mutation factor $F = 0.5$ and crossover rate $C = 0.5$;

(4) CS: Mutation probability $p_a = 0.25$;

(5) GSA: Initial gravitational constant $G_0 = 100$ and decreasing coefficient $a = 20$;

(6) ABC: Limit parameter $= n \cdot d$.

6.3.2 Analysis of exploitation capability

The optimization results offered by the employed optimizers on unimodal functions are described in Table 6.1, in which "Mean" indicates the mean of best-so-far solutions and "Std" indicates the standard deviation of best-so-far solutions. As shown in Table 6.1, SDO provides highly competitive solutions in terms of the "Mean" and "Std" indexes than all other competitors. These results show that SDO is more effective than its competitors in finding the best optimal solutions, demonstrating its superior search ability in terms of exploitation. This merit results from the exploitation mechanism in SDO previously discussed (Zhao et al., 2019).

6.3.3 Analysis of exploration capability

Differing from unimodal functions, both multimodal and fixed-dimensional functions are able to be used to evaluate exploration of algorithms. The optimization results offered by different optimizers on multimodal and fixed-dimensional functions are reported in Tables 6.2 and 6.3, respectively. From these tables, SDO can obtain the best results of all the investigated algorithms for all the functions except functions f_{12}, f_{13}, and f_{20}. However, SDO ranks only second to GA on both functions f_{12} and f_{13}, as well as performs better than PSO, GA, and DE on function f_{20}. The results reveal that SDO shows a distinct advantage in terms of exploration. This is owing to the fact that the exploration mechanism is integrated into SDO (Zhao et al., 2019).

6.3.4 Analysis of avoidance of local optima

It is very difficult for algorithms to find the optimal solutions of composite functions because they are required to balance between exploration and exploitation. Therefore, composite functions are very suitable for assessing local optima avoidance which is able to balance exploration and exploitation. Table 6.4 shows the optimization results of the abovementioned algorithms on composite functions. These results proved that SDO algorithm is significantly effective in balancing exploration and exploitation. Such a merit originated from an adaptive mechanism is employed to the update of search agents in SDO algorithm: The early iterations tend to be dedicated to exploration ($|L| > 1$) while the later iterations to exploitation ($|L| < 1$) (Zhao et al., 2019).

Table 6.1 Result comparisons of algorithms for unimodal functions.

Function	Index	SDO	PSO	GA	DE	CS	GSA	ABC
f_1	Mean	**2.526E − 162**	0.000215	0.009247	3.636E − 14	0.009675	2.195E − 17	0.002363
	Std	7.555E − 162	0.000225	0.003778	6.064E − 14	0.004518	6.379E − 18	0.001525
f_2	Mean	**4.714E − 68**	0.000296	0.021264	4.385E − 08	1.403794	2.283E − 08	2.317E − 04
	Std	5.156E − 68	0.000231	0.005609	2.530E − 08	0.560948	3.490E − 09	1.530E − 04
f_3	Mean	**7.192E − 130**	2.844E + 03	1.053E + 03	5.693864	4.725E + 02	2.216E + 02	9.562E + 03
	Std	3.189E − 130	1.343E + 03	3.489E + 02	3.909791	1.096E + 02	70.667942	1.750E + 03
f_4	Mean	**7.152E − 77**	17.363071	1.046239	9.171683	3.247527	3.451E − 09	24.533250
	Std	2.495E − 77	3.623843	0.289725	3.998583	0.854660	7.445E − 10	2.284116
f_5	Mean	**25.501782**	94.730855	98.563643	3.000E + 01	38.611217	26.694542	5.470E + 02
	Std	0.302722	78.960299	58.445847	1.765E + 01	10.287210	2.671336	2.099E + 02
f_6	Mean	**0**	0.133333	0.000000	0.133333	0.000000	0.000000	0.000000
	Std	0	0.434172	0.000000	0.434172	0.000000	0.000000	0.000000
f_7	Mean	**1.435E − 04**	0.056438	0.044730	0.214589	0.030850	0.019135	0.095282
	Std	8.216E − 05	0.020323	0.013553	0.072382	0.007932	0.006870	0.023852

Table 6.2 Result comparisons of algorithms for multimodal functions.

Function	Index	SDO	PSO	GA	DE	CS	GSA	ABC
f_8	Mean	**−8.740E+03**	−5.139E+03	−6.820E+03	−5.310E+03	−8.692E+03	−2.639E+03	−5.124E+03
	Std	6.856E+02	5.777E+02	5.769E+02	6.617E+02	2.352E+02	4.351E+02	4.607E+02
f_9	Mean	**0**	30.875576	12.400190	1.650E+02	83.232445	15.853010	1.578E+02
	Std	0	8.883905	2.874732	17.511476	13.099442	3.848434	21.302371
f_{10}	Mean	**8.882E−16**	0.007599	0.020177	5.413E−08	4.152485	3.443E−09	0.053290
	Std	0	0.008229	0.004771	2.618E−08	1.486545	5.587E−10	0.040472
f_{11}	Mean	**0**	0.014768	0.021579	0.002054	0.096258	4.264E+00	0.163104
	Std	0	0.013605	0.010212	0.003888	0.041825	1.587503	0.115192
f_{12}	Mean	2.568E−04	0.375773	**2.977E−05**	0.006911	1.100793	0.033992	15.014029
	Std	8.839E−06	0.725728	3.014E−05	0.026302	0.317227	0.053609	4.671204
f_{13}	Mean	8.455E−03	0.190921	**6.310E−04**	0.053060	0.134335	2.041E−18	36.258923
	Std	5.534E−03	0.387922	4.234E−04	0.288556	0.063100	5.131E−19	18.185925

Table 6.3 Result comparisons of algorithms for low-dimensional functions.

Function	Index	SDO	PSO	GA	DE	CS	GSA	ABC
f_{14}	Mean	**0.998004**	0.998004	3.854855	**0.998004**	**0.998004**	3.472815	0.998005
	Std	0	0.000000	2.766177	0	0	2.529209	6.916E − 06
f_{15}	Mean	**3.075E − 04**	3.666E − 04	0.001175	**3.075E − 04**	3.075E − 04	0.002358	5.097E − 04
	Std	9.553E − 20	1.908E − 04	0.001501	1.420E − 19	3.917E − 08	0.001143	4.988E − 05
f_{16}	Mean	**−1.031628**	**−1.031628**	**−1.031628**	**−1.031628**	**−1.031628**	**−1.031628**	**−1.031628**
	Std	2.278E − 16	6.775E − 16	4.277E − 10	6.775E − 16	6.775E − 16	5.532E − 16	6.649E − 16
f_{17}	Mean	**0.397887**	**0.397887**	**0.397887**	**0.397887**	**0.397887**	**0.397887**	**0.397887**
	Std	0	0	1.225E − 08	0	0	0	6.751E − 10
f_{18}	Mean	**3.000000**	**3.000000**	**3.000000**	**3.000000**	**3.000000**	**3.000000**	**3.000000**
	Std	6.027E − 16	2.188E − 15	1.837E − 07	2.030E − 15	1.882E − 15	1.690E − 15	2.861E − 11
f_{19}	Mean	**−3.862782**	**−3.862782**	**−3.862782**	**−3.862782**	**−3.862782**	**−3.862782**	**−3.862782**
	Std	2.278E − 15	2.710E − 15	1.454E − 09	2.710E − 15	2.710E − 15	2.449E − 15	1.668E − 15
f_{20}	Mean	−3.31085	−3.254072	−3.298217	−3.286327	**−3.321995**	**−3.321995**	**−3.321995**
	Std	4.886E − 16	0.060415	0.048370	0.055415	1.917E − 13	1.355E − 15	1.525E − 15
f_{21}	Mean	**−10.153200**	−6.131172	−6.993384	−9.984785	−1.015E + 01	−6.975879	−10.110816
	Std	3.645E − 15	2.835887	3.699551	0.922443	3.803E − 14	3.548979	0.168284
f_{22}	Mean	**−10.402941**	−8.217934	−9.225009	**−10.402941**	**−10.402941**	**−10.402941**	**−10.402941**
	Std	2.967E − 15	2.978914	2.686254	1.745E − 15	2.795E − 14	0.000000	3.529E − 14
f_{23}	Mean	**−10.536410**	−7.716923	−9.340499	**−10.536410**	**−10.536410**	**−10.536410**	**−10.536410**
	Std	1.954E − 15	3.250563	2.731154	1.807E − 15	4.200E − 12	1.682E − 15	8.095E − 14

Table 6.4 Result comparisons of algorithms for composition functions.

Function	Index	SDO	PSO	GA	DE	CS	GSA	ABC
f_{24}	Mean	8.699E−11	98.536796	80.000017	3.298E−04	0.322252	**1.420E−17**	16.358285
	Std	1.111E−10	101.571907	44.721368	0.0152988	0.577046	5.439E−18	35.404507
f_{25}	Mean	**6.362105**	126.37533	199.33855	22.009693	10.208335	200	105.13300
	Std	4.069897	36.780146	90.539010	8.3230419	2.951655	47.140452	28.048742
f_{26}	Mean	**164.847875**	327.81687	476.32103	184.88562	199.36633	172.36085	281.55826
	Std	57.3911974	55.592703	91.317600	34.802273	39.637477	82.861637	48.081802
f_{27}	Mean	**299.384752**	438.28038	577.10983	349.45830	329.10352	333.620196	331.11361
	Std	34.203724	25.953010	43.893813	140.06337	13.329504	154.634938	20.376096
f_{28}	Mean	**2.204867**	60.922190	211.919849	12.579056	10.458422	225.066682	21.134341
	Std	1.845317	37.765887	106.208138	3.4039261	1.953953	86.433716	12.409218
f_{29}	Mean	**500.150184**	636.57241	822.50299	512.06662	505.75894	772.80849	527.71376
	Std	0.194042	184.41463	179.61784	44.758834	1.274708	65.389633	24.504563

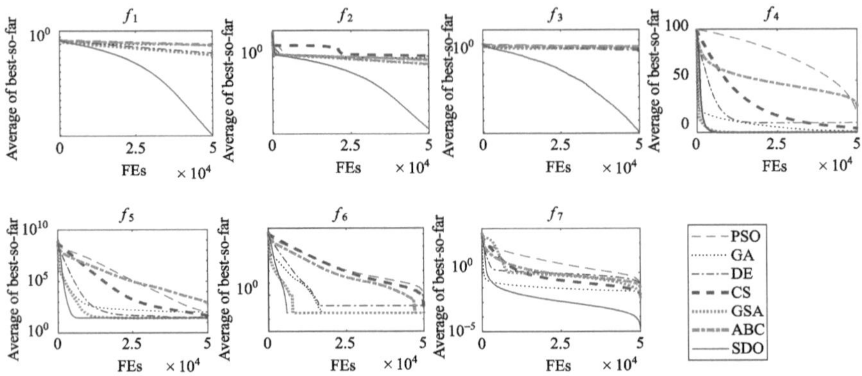

Fig. 6.8 Convergence comparisons of algorithms for unimodal functions.

6.3.5 Analysis of convergence behavior

Figs. 6.8–6.10 show the convergence curves of SDO and other algorithms on different types of functions. Obviously, SDO is very competitive with other metaheuristic methods with respect to convergence rate. SDO shows two different convergence behaviors (Mirjalili, 2016). First, SDO is subject to be accelerated quickly from the initial stage of iterations to the final stage of iterations and then converges to the global optimum (i.e., f_1, f_2, f_3). This is owing to that SDO can find the region with the optimal solution in the early stage of iterations and has an excellent local search ability. Second, SDO tends to converges towards the optimal solution only in the later stage, and this behavior probably results from where SDO fails to find the promising region in the early stage but after multiple attempts it finds the promising region with success in the second half of iterations (i.e., f_8, f_{12}, f_{15}).

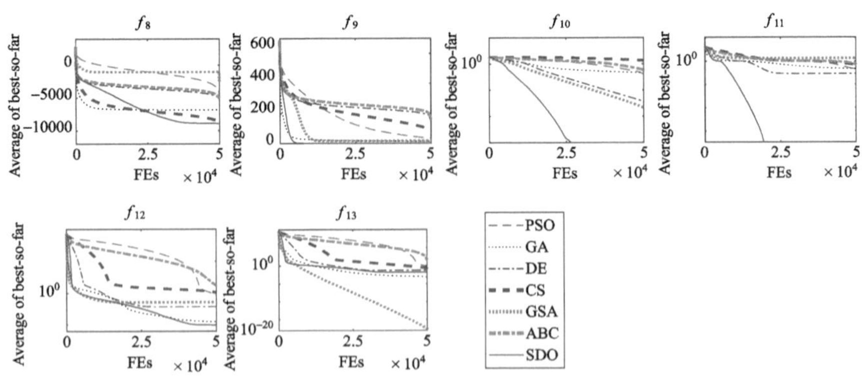

Fig. 6.9 Convergence comparisons of algorithms for multimodal functions.

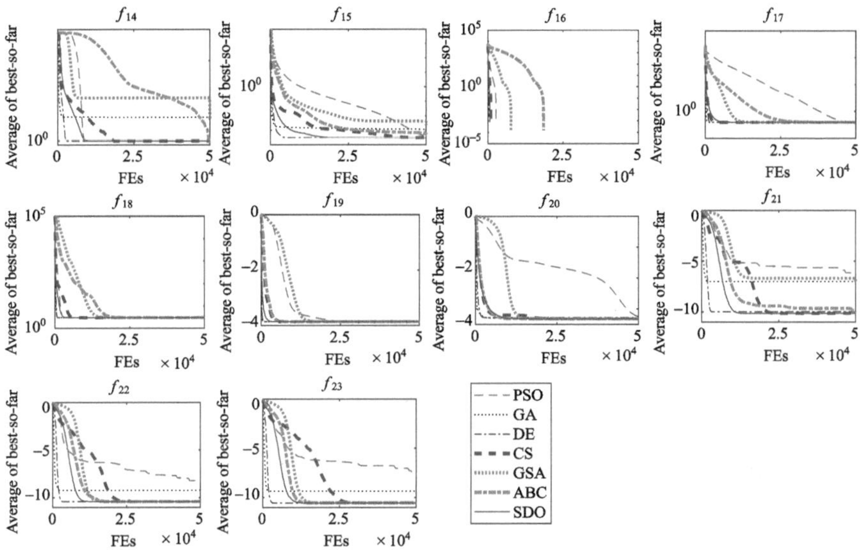

Fig. 6.10 Convergence comparisons of algorithms for low-dimensional functions.

These results denote that SDO has a good ability to explore the global optimum and avoid the local optima. Evidently, these convergence curves reveal that SDO tends to adopt an adaptive search mechanism to different functions, demonstrating its better success in converging to the global optima for different mathematical problems (Zhao et al., 2019).

6.4 Conclusions

The above study shows that SDO has an advantage over other metaheuristic methods. The experimental results on a range of unconstrained benchmark functions reveal the effectiveness of SDO in terms of exploration, exploitation, local optima avoidance, and convergence rate.

References

Chana, I., 2013. Bacterial foraging based hyper-heuristic for resource scheduling in grid computing. Futur. Gener. Comput. Syst. 29, 751–762.

Digalakis, J., Margaritis, K., 2011. On benchmarking functions for genetic algorithms. Int. J. Comput. Math. 77, 481–506.

Ezekiel, M., 1938. The cobweb theorem. Q. J. Econ. 52 (2), 255–280.

Goldstein, A.A., 1965. On steepest descent. J. Soc. Indust. Appl. Math. Ser. A Control 3 (1), 147–151.

Hung, C.L., Wu, Y.H., 2017. Parallel genetic-based algorithm on multiple embedded graphic processing units for brain magnetic resonance imaging segmentation. Comput. Electr. Eng. 61, 373–383.

Ibrahima, H.E.A., Hassanb, F.N., Shomerc, A.O., 2014. Optimal PID control of a brushless dc motor using PSO and BF techniques. Ain Shams Eng. J. 5, 391–398.

Kirkpatrick, S., Gelatt, C.D., Vecchi, M.P., 1983. Optimization by simulated annealing. Science 220 (4598), 671–680.

Lai, C., Shao, Q., Chen, X., Wang, Z., 2016. Flood risk zoning using a rule mining based on ant colony algorithm. J. Hydrol. 542, 268–280.

Liang, J.J., Qu, B.Y., Suganthan, P.N., et al., 2014. Problem Definitions and Evaluation Criteria for the CEC 2014 Special Session and Competition on Single Objective Real-Parameter Numerical Optimization. Computational Intelligence Laboratory, Zhengzhou University, Zhengzhou China and Technical Report, Nanyang Technological University, Singapore.

Maddala, J., Rengaswamy, R., 2012. A genetic algorithm (GA) based rational approach for design of discrete microfluidic networks. Comput. Aided Chem. Eng. 30, 507–511.

Mirjalili, S., 2016. SCA: a sine cosine algorithm for solving optimization problems. Knowl.-Based Syst. 96, 120–133.

Nowcki, H., 1974. Optimization in pre-contract ship design. In: Fujita, Y., Lind, K., Williams, T.J. (Eds.), Computer Applications in the Automation of Shipyard Operation and Ship Design. vol. 2. NorthHolland, Elsevier, New York, pp. 327–338.

Talbi, E.G., 2009. Metaheuristics: From Design to Implementation. vol. 74 John Wiley & Sons, New York.

Wang, L., Zhao, W., Tian, Y., Pan, G., 2018. A bare bones bacterial foraging optimization algorithm. Cogn. Syst. Res. 52, 301–311.

Waugh, F.V., 1964. Cobweb models. Am. J. Agric. Econ. 46 (4), 732–750.

Weerakoon, S., Fernando, T.G.I., 2000. A variant of newton's method with accelerated third-order convergence. Appl. Math. Lett. 13 (8), 87–93.

Zhao, W., Wang, L., 2016. An effective bacterial foraging optimizer for global optimization. Inf. Sci. 329, 719–735.

Zhao, W., Wang, L., Zhang, Z., 2019. Supply–demand-based optimization: a novel economics-inspired algorithm for global optimization. IEEE Access 7, 73182–73206.

CHAPTER 7

Engineering applications of supply-demand-based optimization

Contents

7.1 Introduction

To further verify the performance of the supply-demand-based optimization (SDO) algorithm, five constrained engineering problems are employed in this chapter. These constrained engineering cases are a three-bar truss design, cantilever beam design, rolling element bearing design, and gear train design. They have different constraints with different natures, thus a constraint handling method needs to be utilized. There are different constraint handling methods: static penalty, dynamic penalty, adaptive penalty, annealing penalty, coevolutionary penalty, and death penalty. Of those, penalty function is most frequently used because of its simplicity and lower computational costs. Additionally, this method is not subject to making use of the information of infeasible solutions. Therefore, the SDO algorithm is equipped with the death penalty function to handle constraints in these engineering problems.

7.2 Three-bar truss design

This case (Nowcki, 1974) depicted in Fig. 7.1 deals with a statically loaded three-bar truss with the minimum weight. It has three constraints on stress,

New Optimization Algorithms and their Applications
https://doi.org/10.1016/B978-0-323-90941-9.00007-7

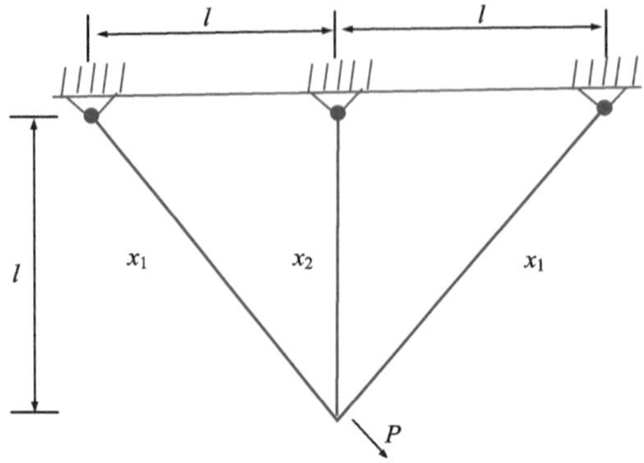

Fig. 7.1 Three-bar truss design problem.

deflection, and buckling and two variables (x_1 and x_2) to adjust the sectional areas. This case is very popular owing to its difficult constrained variable space. The optimization equation for this problem is given in Appendix B.6.

This case was employed by many scholars with some metaheuristic methods in literature: SC (Ray and Liew, 2003), PSO-DE (Liu et al., 2010), DEDS (Zhang et al., 2008), HEAA (Wang et al., 2009), and CS (Gandomi et al., 2013). Our method is compared with these works and the comparisons of their results are provided in Table 7.1. The results of these algorithms report that SDO provides very competitive results in terms of different indices with the same or less computational costs compared to the other algorithms. The comparisons of the best solutions offered by considered methods are provided in Table 7.2. SDO is able to find the best design of all the considered optimizers. Also, it can be seen that although the weights of three-bar truss offered by SDO, PSO-DE, DEDS, and HEAA are equal, the obtained design variables among them are different. Therefore, this suggests that our algorithm can provide a new best optimal design for this case. We give the function value and each constraint value with

Table 7.1 Result comparisons in literature for a three-bar truss design.

Methods	Worst	Mean	Best	Std	FEs
SC	263.969756	263.903356	263.895846	1.3E−02	17,610
PSO−DE	263.895843	263.895843	263.895843	4.5E−10	17,600
DEDS	263.895849	263.895843	263.895843	9.7E−07	15,000
HEAA	263.896099	263.895865	263.895843	4.9E−05	15,000
CS	NA	264.0669	263.97156	9.0E−05	15,000
SDO	263.895847	263.895845	263.895843	1.12E−06	15,000

Table 7.2 Comparison of best solutions offered by reported optimizers for a three-bar truss design.

	SC	PSO-DE	DEDS	HEAA	CS	SDO
x_1	0.788621	0.7886751	0.788675	0.788680	0.788670	0.788698
x_2	0.408401	0.4082482	0.408248	0.408234	0.409020	0.408184
g_1	NA	−5.29E−11	−1.77E−08	NA	−0.00029	−8.052E−10
g_2	NA	−1.4637475	−1.464101	NA	−0.26853	−1.464174
g_3	NA	−0.5362524	−0.535898	NA	−0.73176	−0.535825
f_1	263.895847	263.895843	263.895843	263.895843	263.971623	263.895843

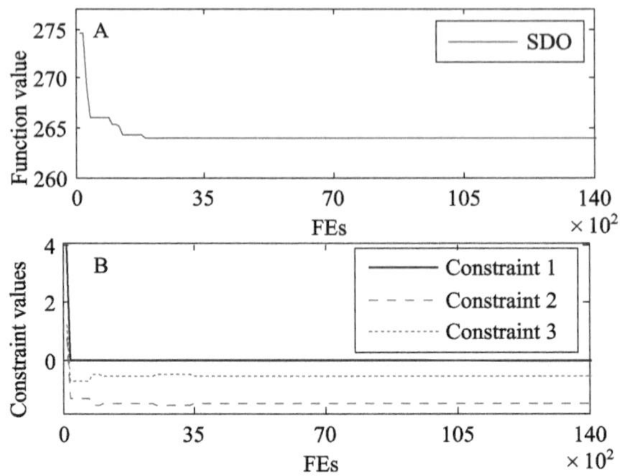

Fig. 7.2 Function value and constraint values versus FEs for a three-bar truss design.

respect to FEs in Fig. 7.2, in which SDO offers a good convergence rate for this problem (Zhao et al., 2019).

7.3 Cantilever beam design

As illustrated in Fig. 7.3, the cantilever beam is composed of five hollow square blocks with constant thickness (Chickermane and Gea, 1996). The beam is rigidly supported at the first block and there is vertical force acting at the free end of the fifth block. So the objective of this case is to minimize the weight of the beam and meanwhile only meet the constraint requirement on an upper limit on the vertical displacement of the free end. There are five decision variables that are respectively lengths of the different blocks. The optimization equation for this problem is given in Appendix B.7.

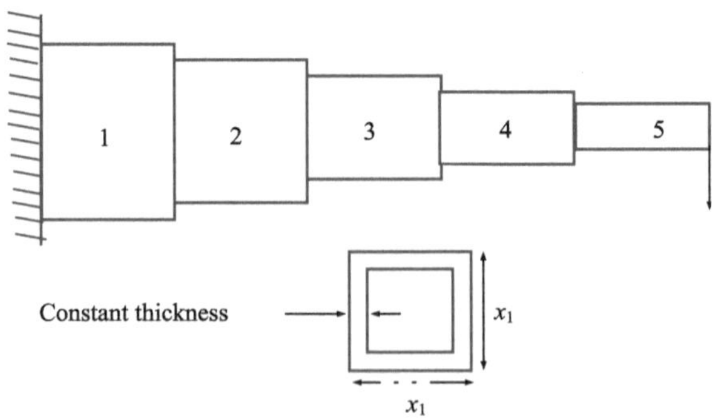

Fig. 7.3 Cantilever beam design problem.

Table 7.3 Result comparisons in the literature for a cantilever beam design.

Methods	Worst	Mean	Best	Std	FEs
SOS	NA	1.33997	1.33996	1.1E−05	15,000
CS	NA	NA	1.33999	NA	125,000
MMA	NA	NA	1.3400	NA	NA
GCA-I	NA	NA	1.3400	NA	NA
GCA-II	NA	NA	1.3400	NA	NA
MFO	NA	NA	3.399880	NA	NA
SDO	1.340513	1.34009	1.339963	2.626E−05	15,000

This problem is attempted using our method and many reported metaheuristics, including SOS (Cheng and Prayogo, 2014), CS (Gandomi et al., 2013), MMA (Chickermane and Gea, 1996), GCA-I (Chickermane and Gea, 1996), GCA-II (Chickermane and Gea, 1996), and MFO (Mirjalili, 2015), and the comparisons of their results are provided in Table 7.3. Overall, the results of SDO have an obvious advantage. In terms of the "Best" index, SDO provides the best results compared to the others. Meanwhile, the results provided by SDO are very close to those provided by SOS in terms of the "Mean" index. The comparisons of the best solutions offered by reported algorithms are provided in Table 7.4. From this table, the results reveal that SDO is able to obtain the best optimal design for this problem. Fig. 7.4 shows the function value and constraint value with respect to FEs.

7.4 Rolling element bearing design

This problem is to maximize the dynamic load carrying capacity of rolling element bearing as illustrated in Fig. 7.5. The maximization process is subject to constraints on the geometry and kinematics. This problem has

Table 7.4 Comparisons of best solutions offered by reported optimizers for a cantilever beam design.

	SOS	CS	MMA	GCA-I	GCA-II	MFO	SDO
x_1	6.01878	6.0089	6.0100	6.0100	6.0100	5.984871	6.021609
x_2	5.30344	5.3049	5.3000	5.3000	5.3000	5.316726	5.294696
x_3	4.49587	4.5023	4.4900	4.4900	4.4900	4.497332	4.500152
x_4	3.49896	3.5077	3.4900	3.4900	3.4900	3.513616	3.505450
x_5	2.15564	2.1504	2.1500	2.1500	2.1500	2.1616200	2.151867
g_1	NA	NA	NA	NA	NA	NA	−1.547E−9
f_2	1.33996	1.33999	1.3400	1.3400	1.3400	1.339988	1.339963

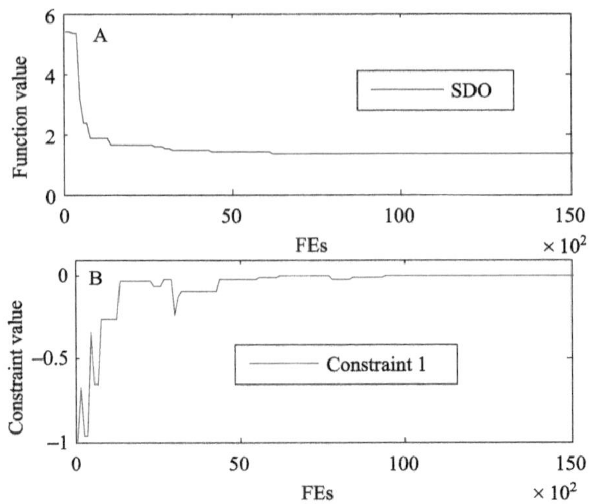

Fig. 7.4 Function value and constraint value versus FEs for a cantilever beam design.

10 decision variables: pitch diameter (D_m), ball diameter (D_b), number of balls (Z), inner and outer raceway curvature coefficients (f_i and f_o), and other internal geometry parameters ($K_{D\min}$, $K_{D\max}$, ε, e, and ζ) that only appear in constraints. The optimization equation for this problem is given in Appendix B.8.

The problem was attempted by GA4 (Rao and Tiwari, 2007), TLBO (Gupta et al., 2017), ABC (Rao et al., 2011), and MBA (Sadollah et al., 2013); the best results of these approaches and SDO are compared in Table 7.5. It can be seen that the proposed method is able to find the best

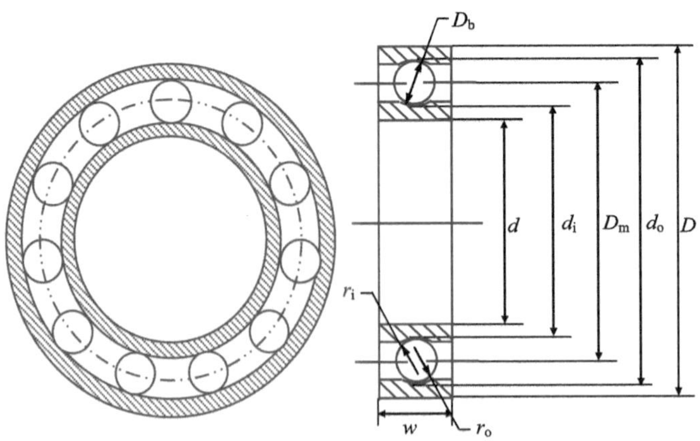

Fig. 7.5 Rolling element bearing design.

Table 7.5 Result comparisons in literature for a rolling element bearing design.

Methods	Worst	Mean	Best	Std	FEs
GA4	NA	NA	81,843.3000	NA	225,000
ABC	78,897.8100	81,496.0000	81,859.7416	NA	10,000
TLBO	80,807.8551	81,438.9870	81,859.7400	NA	10,000
MBA	84,440.1948	85,321.4030	85,535.9611	211.5200	15,100
SDO	85,144.1639	85,443.6665	85,538.85468	112.2171	15,000

optimal solutions in terms of the "Worst," "Mean," "Best," "Std" indexes over the other algorithms. Although the number of function evaluations of SDO is more than those of both ABC and TLBO, the overall results that the performance of SDO is considerably better than that of both algorithms for this problem. Table 7.6 shows the comparisons of best optimal solutions for the considered algorithms in terms of design decision variables and function values. This table reveals that SDO obtains a design with the maximum dynamic load-carrying capacity. Fig. 7.6 shows the function value and each constraint value with respect to FEs (Zhao et al., 2019).

Table 7.6 Comparisons of best solutions offered by reported optimizers for a rolling element bearing design.

	GA4	TLBO	MBA	SDO
$x_1(D_m)$	125.7171	125.7191	125.7153	1.257E+02
$x_2(D_b)$	21.423	21.42559	21.423300	21.424905
$x_3(Z)$	11	11	11.000	11.430211
$x_4(f_i)$	0.515	0.515	0.515000	0.515002
$x_5(f_o)$	0.515	0.515	0.515000	0.519304
$x_6(K_{D\min})$	0.4159	0.424266	0.488805	0.487755
$x_7(K_{D\max})$	0.651	0.633948	0.627829	0.629992
$x_8\ (\varepsilon)$	0.300043	0.3	0.300149	0.300039
$x_9(e)$	0.0223	0.068858	0.097305	5.351E−02
$x_{10}(\zeta)$	0.751	0.799498	0.646095	0.665982
g_1	0.000821	0	0	1.107E−04
g_2	13.732999	13.15257	8.630183	8.706977
g_3	2.724000	1.5252	1.101429	1.249638
g_4	3.606000	0.719056	2.040448	14.094788
g_5	0.717000	16.49544	0.715366	12.660371
g_6	4.857899	0	23.611002	0.717208
g_7	0.003050	0	0.000480	6.393E−04
g_8	0.000007	2.559363	0	1.445455
g_9	0.000007	0	0	2.076E−06
g_{10}	0.000005	0	0	4.304E−03
f_3	81,843.3	81,859.74	85,535.9611	85,538.85468

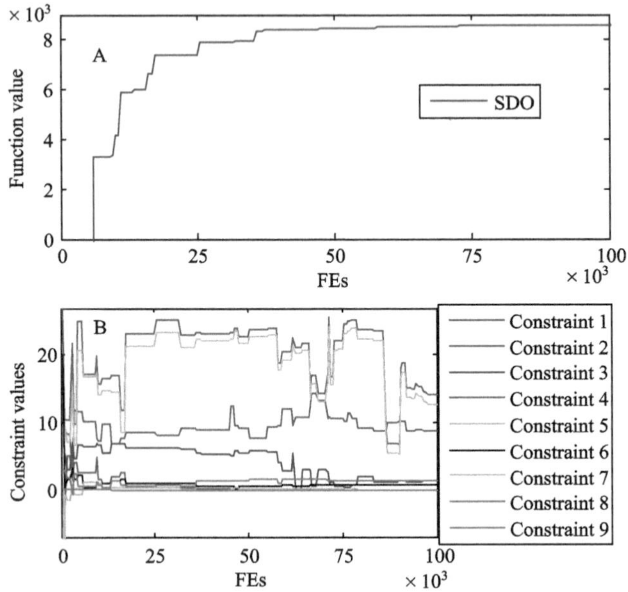

Fig. 7.6 Function value and constraint values versus FEs for rolling element bearing design.

7.5 Gear train design

Gear train design aims to minimize the error between the obtained gear ratio and a required gear ratio of 1/6.931. The constraints are only limits on design variables (side constraints). Design variables to be optimized are in a discrete form because each gear has to have an integral number of teeth. The design variables T_a, T_b, T_d, and T_f represent the teeth number of four gears, respectively. The lower and upper bounds of integer design variables are 12 and 60, respectively (Sadollah et al., 2013). The gear train design is shown in Fig. 7.7. The optimization equation for this problem is given in Appendix B.9.

Table 7.7 shows the comparison of the best solution for SDO and other algorithms such MFO, MBA, and PSO in terms of the value of design variables and function value. It can be seen that, although the error between the obtained gear ratio and a required gear ratio provided by AEO is the same as that provided by MBA and MFO, the optimal design variables obtained are different. Therefore, the results evidence that AEO provides a good alternative to solve this problem. Fig. 7.8 shows objective function value, design variables versus iterations provided by the SDO for gear train design problem.

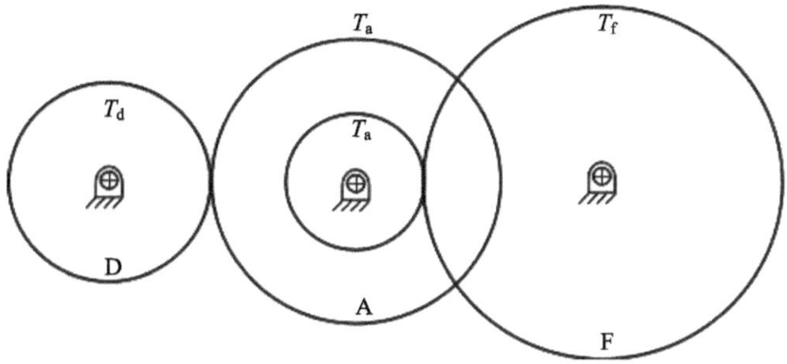

Fig. 7.7 Gear train design problem.

Table 7.7 Values of objective functions, design variables for gear train design problem.

	Optimal variables				
Algorithms	T_a	T_b	T_d	T_f	**Function value**
MFO	43	19	16	49	2.7009E−12
SDO	49	16	19	43	2.7009E−12
MBA	43	16	19	49	2.7009E−12
PSO	52	15	29	58	2.35764E−9

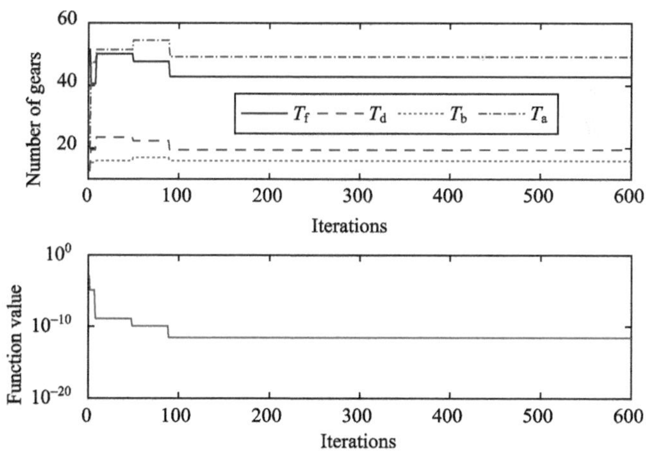

Fig. 7.8 Design variables versus iterations, value of objective function.

7.6 Conclusions

The comparisons using four classical constrained engineering problems demonstrate the competiveness of our algorithm in terms of computational expense and convergence rate, highlighting its potential in solving challenging real-world problems.

SDO provides enough competitive results compared to other considered metaheuristic methods based on the reported results in the literature and the experimental results in this study. For future work, the binary SDO may be presented to deal with complex discrete problems. SDO could also be tailored to address multiobjective optimization.

References

Cheng, M.Y., Prayogo, D., 2014. Symbiotic organisms search: a new metaheuristic optimization algorithm. Comput. Struct. 139, 98–112.

Chickermane, H., Gea, H.C., 1996. Structural optimization using a new local approximation method. Int. J. Numer. Methods Eng. 39 (5), 829–846.

Gandomi, A.H., Yang, X.S., Alavi, A.H., et al., 2013. Cuckoo search algorithm: a metaheuristic approach to solve structural optimization problems. Eng. Comput. 29 (1), 17–35.

Gupta, S., Tiwari, R., Shivashankar, B.N., et al., 2017. Multi-objective design optimization of rolling bearings using genetic algorithm. Mech. Mach. Theory 42, 1418–1443.

Liu, H., Cai, Z., Wang, Y., 2010. Hybridizing particle swarm optimization with differential evolution for constrained numerical and engineering optimization. Appl. Soft Comput. 10, 629–640.

Mirjalili, S., 2015. Moth-flame optimization algorithm: a novel nature-inspired heuristic paradigm. Knowl.-Based Syst. 89, 228–249.

Nowcki, H., 1974. Optimization in pre-contract ship design. In: Fujita, Y., Lind, K., Williams, T.J. (Eds.), Computer Applications in the Automation of Shipyard Operation and Ship Design. In: vol. 2. NorthHolland, Elsevier, New York, pp. 327–338.

Rao, B.R., Tiwari, R., 2007. Optimum design of rolling element bearings using genetic algorithms. Mech. Mach. Theory 42 (2), 233–250.

Rao, R.V., Savsani, V.J., Vakharia, D.P., et al., 2011. Teaching-learning-based optimization: a novel method for constrained mechanical design optimization problems. Comput. Aided Des. 43 (3), 303–315.

Ray, T., Liew, K.M., 2003. Society and civilization: an optimization algorithm based on the simulation of social behavior. IEEE Trans. Evol. Comput. 7 (4), 386–396.

Sadollah, A., Bahreininejad, A., Eskandar, H., et al., 2013. Mine blast algorithm: a new population based algorithm for solving constrained engineering optimization problems. Appl. Soft. Comput. 13 (5), 2592–2612.

Wang, Y., Cai, Z., Zhou, Y., Fan, Z., 2009. Constrained optimization based on hybrid evolutionary algorithm and adaptive constraint handling technique. Struct. Multidiscip. Optim. 37, 395–413.

Zhang, M., Luo, W., Wang, X., 2008. Differential evolution with dynamic stochastic selection for constrained optimization. Inf. Sci. 178, 3043–3074.

Zhao, W., Wang, L., Zhang, Z., 2019. Supply-demand-based optimization: a novel economics-inspired algorithm for global optimization. IEEE Access 7, 73182–73206.

Appendix

Appendix A Benchmark functions

Table A.1 Unimodal test functions.

Name	Function	D	Range	f_{opt}
Sphere	$f_1(x) = \sum_{i=1}^{n} x_i^2$	30	$[-100, 100]^D$	0
Schwefel 2.22	$f_2(x) = \sum_{i=1}^{n} \lvert x_i \rvert + \prod_{i=1}^{n} \lvert x_i \rvert$	30	$[-10, 10]^D$	0
Schwefel 1.2	$f_3(x) = \sum_{i=1}^{n} \left(\sum_{j=1}^{i} x_j \right)^2$	30	$[-100, 100]^D$	0
Schwefel 2.21	$f_4(x) = \max_i \{ \lvert x_i \rvert, 1 \le i \le n \}$	30	$[-100, 100]^D$	0
Rosenbrock	$f_5(x) = \sum_{i=1}^{n-1} \left[100(x_{i+1} - x_i)^2) + (x_i - 1)^2 \right]$	30	$[-30, 30]^D$	0
Step	$f_6(x) = \sum_{i=1}^{n} (x_i + 0.5)^2$	30	$[-100, 100]^D$	0
Quartic	$f_7(x) = \sum_{i=1}^{n} i x_i^4 + \text{rand}[0, 1)$	30	$[-1.28, 1.28]^D$	0

Table A.2 Multimodal test functions.

Name	Function	D	Range	f_{opt}
Schwefel	$f_8(x) = -\sum_{i=1}^{n}\left[x_i \sin\left(\sqrt{\lvert x_i \rvert}\right)\right]$	30	$[-500, 500]^D$	−12,569.5
Rastrigin	$f_9(x) = \sum_{i=1}^{n}\left[x_i^2 - 10\cos(2\pi x_i) + 10\right]^2$	30	$[-5.12, 5.12]^D$	0
Ackley	$f_{10}(x) = -20\exp\left(-0.2\sqrt{\frac{1}{n}\sum_{i=1}^{n} x_i^2}\right) - \exp\left(\frac{1}{n}\sum_{i=1}^{n}\cos 2\pi x_i\right) + 20 + e$	30	$[-32, 32]^D$	0
Griewank	$f_{11}(x) = \frac{1}{4000}\sum_{i=1}^{n}(x_i - 100)^2 - \prod_{i=1}^{n}\cos\left(\frac{x_i - 100}{\sqrt{i}}\right) + 1$	30	$[-600, 600]^D$	0
Penalized	$f_{12}(x) = \frac{\pi}{n} 10\sin^2(\pi y_1) + \sum_{i=1}^{n-1}(y_i - 1)^2\left[1 + 10\sin^2(\pi y_i + 1)\right] + (y_n - 1)^2 + \sum_{i=1}^{30} u(x_i, 10, 100, 4)$	30	$[-50, 50]^D$	0
Penalized2	$f_{13}(x) = 0.1\sin^2(3\pi x_1) + \sum_{i=1}^{29}(x_i - 1)^2 p\left[1 + \sin^2(3\pi x_{i+1})\right] + (x_n - 1)^2\left[1 + \sin^2(2\pi x_{30})\right] + \sum_{i=1}^{30} u(x_i, 5, 10, 4)$	30	$[-50, 50]^D$	0

Table A.3 Low-dimensional multimodal test functions.

Name	Function	D	Range	f_{opt}
Foxholes	$f_{14}(x)=\left[\frac{1}{500}+\sum_{j=1}^{25}\frac{1}{j+\sum_{j=1}^{2}(x_i-a_{ij})^6}\right]^{-1}$	2	$[-65.536, 65.536]^D$	0.998
Kowalik	$f_{15}(x)=\sum_{i=1}^{11}\left\|a_i-\frac{x_1(b_i^2+b_ix_2)}{b_i^2+b_ix_3+x_4}\right\|^2$	4	$[-5, 5]^D$	3.075×10^{-4}
Six Hump Camel	$f_{16}(x)=4x_1^2-2.1x_1^4+\frac{1}{3}x_1^6+x_1x_2-4x_2^2+4x_2^4$	2	$[-5, 5]^D$	−1.0316
Branin	$f_{17}(x)=\left(x_2-\frac{5.1}{4\pi^2}x_1^2+\frac{5}{\pi}x_1-6\right)^2+10\left(1-\frac{1}{8\pi}\right)\cos x_1+10$	2	$[-5, 10]\times[0, 15]$	0.398
Goldstein-Price	$f_{18}(x)=\left[1+(x_1+x_2+1)^2(19-14x_1+3x_1^2-14x_2+6x_1x_2+3x_2^2)\right]\times\left[30+(2x_1+1-3x_2)^2(18-32x_1+12x_1^2+48x_2-36x_1x_2+27x_2^2)\right]$	2	$[-2.2]^D$	3
Hartman 3	$f_{19}(x)=-\sum_{i=1}^{4}\exp\left[-\sum_{j=1}^{3}a_{ij}(x_j-p_{ij})^2\right]$	3	$[0, 1]^D$	−3.86
Hartman 6	$f_{20}(x)=-\sum_{i=1}^{4}\exp\left[-\sum_{j=1}^{6}a_{ij}(x_j-p_{ij})^2\right]$	6	$[0, 1]^D$	−3.322
Shekel 5	$f_{21}(x)=-\sum_{i=1}^{5}\left\|(x_i-a_i)(x_i-a_i)^T+c_i\right\|^{-1}$	4	$[0, 10]^D$	−10.1532
Shekel 7	$f_{22}(x)=-\sum_{i=1}^{7}\left\|(x_i-a_i)(x_i-a_i)^T+c_i\right\|^{-1}$	4	$[0, 10]^D$	−10.4028
Shekel 10	$f_{23}(x)=-\sum_{i=1}^{10}\left\|(x_i-a_i)(x_i-a_i)^T+c_i\right\|^{-1}$	4	$[0, 10]^D$	−10.5364

Table A.4 Composition test functions.

Function	Name	n	Range	Optimum
f_{24}	Hybrid function 1 ($n = 3$)	30	$[-100, 100]^n$	1700
f_{25}	Hybrid function 2 ($n = 3$)	30	$[-100, 100]^n$	1800
f_{26}	Hybrid function 3 ($n = 4$)	30	$[-100, 100]^n$	1900
f_{27}	Hybrid function 4 ($n = 4$)	30	$[-100, 100]^n$	2000
f_{28}	Hybrid function 5 ($n = 5$)	30	$[-100, 100]^n$	2100
f_{29}	Hybrid function 6 ($n = 5$)	30	$[-100, 100]^n$	2200
f_{30}	Composition function 1 ($n = 5$)	30	$[-100, 100]^n$	2300
f_{31}	Composition function 2 ($n = 3$)	30	$[-100, 100]^n$	2400
f_{32}	Composition function 3 ($n = 3$)	30	$[-100, 100]^n$	2500
f_{33}	Composition function 4 ($n = 5$)	30	$[-100, 100]^n$	2600
f_{34}	Composition function 5 ($n = 5$)	30	$[-100, 100]^n$	2700
f_{35}	Composition function 6 ($n = 5$)	30	$[-100, 100]^n$	2800
f_{36}	Composition function 7 ($n = 3$)	30	$[-100, 100]^n$	2900
f_{37}	Composition function 8 ($n = 3$)	30	$[-100, 100]^n$	3000

Table A.5 Hybrid and composition test functions.

Function	Name	D	Range	f_{opt}
f_{24}	Composition function 1 ($n = 5$)	30	$[-100, 100]^D$	2300
f_{25}	Composition function 2 ($n = 3$)	30	$[-100, 100]^D$	2400
f_{26}	Composition function 3 ($n = 3$)	30	$[-100, 100]^D$	2500
f_{27}	Composition function 4 ($n = 5$)	30	$[-100, 100]^D$	2600
f_{28}	Composition function 5 ($n = 5$)	30	$[-100, 100]^D$	2700
f_{29}	Composition function 6 ($n = 5$)	30	$[-100, 100]^D$	2800
f_{30}	Composition function 7 ($n = 3$)	30	$[-100, 100]^D$	2900
f_{31}	Composition function 8 ($n = 3$)	30	$[-100, 100]^D$	3000

Table A.6 Composition test functions.

Name	Function	D	Range	f_{opt}
f_{24} (CF1)	$f_1, f_2, f_3, \cdots, f_{10}$ = Sphere function $[\sigma_1, \sigma_2, \sigma_3, \cdots, \sigma_{10}] = [1, 1, 1, \cdots, 1]$ $[\lambda_1, \lambda_2, \lambda_3, \cdots, \lambda_{10}] = [5/100, 5/100, 5/100, \cdots, 5/100]$	10	$[-5, 5]^D$	0
f_{25} (CF2)	$f_1, f_2, f_3, \cdots, f_{10}$ = Griewank0s function $[\sigma_1, \sigma_2, \sigma_3, \cdots, \sigma_{10}] = [1, 1, 1, \cdots, 1]$ $[\lambda_1, \lambda_2, \lambda_3, \cdots, \lambda_{10}] = [5/100, 5/100, 5/100, \cdots, 5/100]$	10	$[-5, 5]^D$	0

Table A.6 Composition test functions—cont'd

Name	Function	D	Range	f_{opt}
f_{26} (CF3)	$f_1, f_2, f_3, \cdots, f_{10}$ = Griewank0s function $[\sigma_1, \sigma_2, \sigma_3, \cdots, \sigma_{10}] = [1, 1, 1, \cdots, 1]$ $[\lambda_1, \lambda_2, \lambda_3, \cdots, \lambda_{10}] = [1, 1, 1, \cdots, 1]$	10	$[-5, 5]^D$	0
f_{27} (CF4)	f_1, f_2 = Ackley function f_3, f_4 = Rastrigin function f_5, f_6 = Weierstrass function f_7, f_8 = Griewank function f_9, f_{10} = Sphere function $[\sigma_1, \sigma_2, \sigma_3, \cdots, \sigma_{10}] = [1, 1, 1, \cdots, 1]$ $[\lambda_1, \lambda_2, \lambda_3, \cdots, \lambda_{10}] = [5/32, 5/32, 1, 1, 5/0.5, 5/0.5, 5/100, 5/100, 5/100, 5/100]$	10	$[-5, 5]^D$	0
f_{28} (CF5)	f_1, f_2 = Rastrigin0s function f_3, f_4 = Weierstrass function f_5, f_6 = Griewank function f_7, f_8 = Ackley0s function f_9, f_{10} = Sphere function $[\sigma_1, \sigma_2, \sigma_3, \cdots, \sigma_{10}] = [1, 1, 1, \cdots, 1]$ $[\lambda_1, \lambda_2, \lambda_3, \cdots, \lambda_{10}] = [1/5, 1/5, 5/0.5, 5/0.5, 5/100, 5/100, 5/32, 5/32, 5/100, 5/100]$	10	$[-5, 5]^D$	0
f_{29} (CF6)	f_1, f_2 = Rastrigin function f_3, f_4 = Weierstrass function f_5, f_6 = Griewank function f_7, f_8 = Ackley function f_9, f_{10} = Sphere function $[\sigma_1, \sigma_2, \sigma_3, \cdots, \sigma_{10}] = [0.1, 0.2, 0.3, 0.4, 0.5, 0.6, 0.7, 0.8, 0.9, 1]$ $[\lambda_1, \lambda_2, \lambda_3, \cdots, \lambda_{10}] = [0.1 \times 1/5, 0.2 \times 1/5, 0.3 \times 5/0.5, 0.4 \times 5/0.5, 0.5 \times 5/100, 0.6 \times 5/100, 0.7 \times 5/32, 0.8 \times 5/32, 0.9 \times 5/100, 5/100]$	10	$[-5, 5]^D$	0

Appendix B Engineering design problems

B.1 Tension/compression spring design

Consider variable $\boldsymbol{x} = [x_1, x_2, x_3] = [d, D, N]$.

Minimize $f_1(\boldsymbol{x}) = (x_3 + 2)x_2x_1^2$.

Subject to $g_1(\boldsymbol{x}) = 1 - \frac{x_3x_2^3}{71785x_1^4} \le 0$, $g_2(\boldsymbol{x}) = \frac{4x_2^2 - x_1x_2}{12566\left(x_2x_1^3 - x_1^4\right)} + \frac{1}{5108x_1^2} - 1 \le 0$,
$g_3(\boldsymbol{x}) = 1 - \frac{140.45x_1}{x_2^2x_3} \le 0$, $g_4(\boldsymbol{x}) = \frac{x_1 + x_2}{1.5} - 1 \le 0$.

Variable range $0.05 \le x_1 \le 2$, $0.25 \le x_2 \le 1.3$, $2 \le x_3 \le 15$.

B.2 Pressure vessel design

Consider variable $\boldsymbol{x} = [x_1, x_2, x_3, x_4] = [T_s, T_h, R, L]$.

Minimize $f_2(\boldsymbol{x}) = 0.6224x_1x_3x_4 + 1.7781x_2x_3^2 + 3.1661x_1^2x_4 + 19.84x_1^2x_3$.

Subject to $g_1(\boldsymbol{x}) = -x_1 + 0.0193x_3 \leq 0$, $g_2(\boldsymbol{x}) = -x_2 + 0.00954x_3 \leq 0$, $g_3(\boldsymbol{x}) = -\pi x_3^2x_4 - \frac{4}{3}\pi x_3^3 + 1296000 \leq 0$, $g_4(\boldsymbol{x}) = x_4 - 240 \leq 0$.

Variable range $0 \leq x_1 \leq 99$, $0 \leq x_2 \leq 99$, $10 \leq x_3 \leq 200$, $10 \leq x_4 \leq 200$.

B.3 Welded beam design

Consider variable $\boldsymbol{x} = [x_1, x_2, x_3, x_4] = [h, l, t, b]$.

Minimize $f_3(\boldsymbol{x}) = 1.10471x_1^2x_2 + 0.04811x_3x_4(14 + x_2)$.

Subject to $g_1(\boldsymbol{x}) = \tau(\boldsymbol{x}) + \tau_{\max} \leq 0$, $g_2(\boldsymbol{x}) = \sigma(\boldsymbol{x}) + \sigma_{\max} \leq 0$, $g_3(\boldsymbol{x}) = \delta(\boldsymbol{x}) + \delta_{\max} \leq 0$, $g_4(\boldsymbol{x}) = x_1 - x_4 \leq 0$, $g_5(\boldsymbol{x}) = P - P_c(\boldsymbol{x}) \leq 0$, $g_6(\boldsymbol{x}) = 0.125 - x_1 \leq 0$, $g_7(\boldsymbol{x}) = 0.10471x_1^2 + 0.04811x_3x_4(14 + x_2) - 5 \leq 0$.

where

$$\tau(\boldsymbol{x}) = \sqrt{(\tau')^2 + 2\tau'\tau''\frac{x_2}{2R} + (\tau'')^2},\ \tau' = \frac{P}{\sqrt{2}x_1x_2},\ \tau'' = \frac{MR}{J},\ M = P\left(L + \frac{x_2}{2}\right),$$

$$R = \sqrt{\frac{x_2^2}{4} + \left(\frac{x_1 + x_3}{2}\right)^2},\quad J = 2\left\{\sqrt{2}x_1x_2\left[\frac{x_2^3}{4} + \left(\frac{x_1 + x_3}{2}\right)^2\right]\right\},\quad \sigma(\boldsymbol{x}) = \frac{6PL}{x_4x_3^2},$$

$$\delta(\boldsymbol{x}) = \frac{4PL^3}{Ex_4x_3^3},\ P_c(\boldsymbol{x}) = \frac{4.013E\sqrt{\frac{x_3^2x_4^6}{36}}}{L^2}\left(1 - \frac{x_3}{2L}\sqrt{\frac{E}{4G}}\right).$$

where $P = 6000$ lb.,[a] $L = 14$ in.,[b] $E = 30 \times 10^6$psi,[c] $G = 12 \times 10^6$psi, $\tau_{\max} = 13600$ psi, $\sigma_{\max} = 30000$ psi, $\delta_{\max} = 0.25$ in.

Variable range $0.1 \leq x_1 \leq 2$, $0.1 \leq x_2 \leq 10$, $0.1 \leq x_3 \leq 10$, $0.1 \leq x_4 \leq 2$.

B.4 Speed reducer design

Consider variable $\boldsymbol{x} = [x_1, x_2, x_3, x_4, x_5, x_6, x_7] = [b, m, z, l_1, l_2, d_1, d_2]$.

Minimize $f_4(\boldsymbol{x}) = 0.7854x_1x_2^2\ (3.3333x_3^2 + 14.9334x_3 - 43.0934) -1.508x_1(x_6^2 + x_7^2) + 0.7854x_1(x_4x_6^2 - x_5x_7^2)$.

Subject to $g_1(\boldsymbol{x}) = \frac{27}{x_1x_2^2x_3} - 1 \leq 0$, $g_2(\boldsymbol{x}) = \frac{397.5}{x_1x_2^2x_3^2} - 1 \leq 0$, $g_3(\boldsymbol{x}) = \frac{1.93x_4^2}{x_2x_6^4x_3} - 1 \leq 0$,

$$g_4(\boldsymbol{x}) = \frac{1.93x_5^2}{x_2x_7^4x_3} - 1 \leq 0,\ g_5(\boldsymbol{x}) = \frac{\left[\left(\frac{745x_4}{x_2x_3}\right)^2 + 16.9\times10^6\right]^{0.5}}{110x_6^3} - 1 \leq 0,$$

[a] 1 lb. = 0.4535924 kg.

[b] 1 in. = 0.0254 m.

[c] 1 psi = 6.894757 kPa.

$$g_6(\boldsymbol{x}) = \frac{\left[\left(\frac{745x_5}{x_2x_3}\right)^2 + 157.5\times 10^6\right]^{0.5}}{85x_7^3} - 1 \le 0,\quad g_7(\boldsymbol{x}) = \frac{x_2x_3}{40} - 1 \le 0,\quad g_8(\boldsymbol{x}) = \frac{5x_2}{x_1} - 1 \le 0,$$

$$g_9(\boldsymbol{x}) = \frac{x_1}{12x_2} - 1 \le 0, g_{10}(\boldsymbol{x}) = \frac{1.5x_6 + 1.9}{x_4} - 1 \le 0,\ g_{11}(\boldsymbol{x}) = \frac{1.1x_7 + 1.9}{x_5} - 1 \le 0.$$

Variable range $2.6 \le x_1 \le 3.6$, $0.7 \le x_2 \le 0.8$, $17 \le x_3 \le 28$, $7.3 \le x_4 \le 8.3$, $7.3 \le x_5 \le 8.3$, $2.9 \le x_6 \le 3.9$, $5.0 \le x_7 \le 5.5$.

B.5 Multiple disc clutch brake design

Consider variable $\boldsymbol{x} = [r_i, r_o, t, F, Z]$.

Minimize $f_5(\boldsymbol{x}) = \pi(r_o^2 - r_i^2)t(Z+1)\rho$.

Subject to $g_1(\boldsymbol{x}) = r_o - r_i - \Delta r \ge 0$, $g_2(\boldsymbol{x}) = l_{\max} - (Z+1)(t+\delta) \ge 0$, $g_3(\boldsymbol{x}) = p_{\max} - p_{rz} \ge 0$, $g_4(\boldsymbol{x}) = p_{\max}v_{sr\max} - p_{rz}v_{sr} \ge 0$, $g_5(\boldsymbol{x}) = v_{sr\max} - v_{sr} \ge 0$, $g_6(\boldsymbol{x}) = T_{\max} - T \ge 0$, $g_7(\boldsymbol{x}) = M_h - sM_s \ge 0$, $g_8(\boldsymbol{x}) = T \ge 0$.

where

$$M_h = \frac{2}{3}\mu F Z \frac{r_o^3 - r_i^3}{r_o^2 - r_i^2},\quad p_{rz} = \frac{F}{\pi\left(r_o^2 - r_i^2\right)},\quad v_{sr} = \frac{2\pi n\left(r_o^3 - r_i^3\right)}{90\left(r_o^2 - r_i^2\right)},\quad T = \frac{I_z \pi n}{30\left(M_h + M_f\right)},$$

$\Delta r = 20\,\text{mm}$, $l_{\max} = 30\,\text{mm}$, $v_{sr\max} = 10\,\text{m/s}$, $\delta = 0.5$, $s = 1.5$, $M_s = 40\,\text{Nm}$, $M_f = 3\,\text{Nm}$, $n = 250\,\text{rpm}$, $p_{\max} = 1\,\text{MPa}$, $I_z = 55\,\text{kg mm}^2$, $T_{\max} = 15\,\text{s}$, $F_{\max} = 1000\,\text{N}$, $r_{i\min} = 60\,\text{mm}$, $r_{i\max} = 80\,\text{mm}$, $r_{o\min} = 90\,\text{mm}$, $r_{o\max} = 110\,\text{mm}$, $t_{\min} = 1\,\text{mm}$, $t_{\max} = 3\,\text{mm}$, $F_{\min} = 600$, $F_{\max} = 1100$, $Z_{\min} = 2$, $Z_{\max} = 9$.

B.6 Three-bar truss design

Consider variable $\boldsymbol{x} = [x_1, x_2]$.

Minimize $f_6(\mathbf{x}) = \left(2\sqrt{2}x_2 + x_2\right) \times l$.

Subject to $g_1(\boldsymbol{x}) = \frac{\sqrt{2}x_1 + x_2}{\sqrt{2}x_1^2 + 2x_2x_1}P - \sigma \le 0$, $g_2(\boldsymbol{x}) = \frac{x_2}{\sqrt{2}x_1^2 + 2x_2x_1}P - \sigma \le 0$, $g_3(\boldsymbol{x}) = \frac{x_2}{x_1 + \sqrt{2}x_2}P - \sigma \le 0$.

where $l = 10$ cm, $P = 2\,\text{kN/cm}^2$, $\sigma = 2\,\text{kN/cm}^2$.

Variable range $0 \le x_1 \le 1$, $0 \le x_2 \le 1$.

B.7 Cantilever beam design

Consider variable $\boldsymbol{x} = [x_1, x_2, x_3, x_4, x_5]$.

Minimize $f_7(\boldsymbol{x}) = 0.0624(x_1 + x_2 + x_3 + x_4 + x_5)$.

Subject to $g_1(\boldsymbol{x}) = \frac{61}{x_1^3} + \frac{37}{x_2^3} + \frac{19}{x_3^3} + \frac{7}{x_4^3} + \frac{1}{x_5^3} - 1 \le 0$.

Variable range $0.01 \le x_i \le 100$, $i = 1, \cdots, 5$.

B.8 Rolling element bearing design

Consider variable $\boldsymbol{x} = [D_{\mathrm{m}}, D_{\mathrm{b}}, Z, f_{\mathrm{i}}, f_{\mathrm{o}}, K_{D\mathrm{min}}, K_{D\mathrm{max}}, \varepsilon, e, \zeta]$.

$$\text{Maximize} \begin{cases} f_8(\boldsymbol{x}) = f_c Z^{2/3} D_{\mathrm{b}}^{1.8} & \text{if } D_{\mathrm{b}} \leq 25.4\text{mm} \\ f_8(\boldsymbol{x}) = 3.647 f_c Z^{2/3} D_{\mathrm{b}}^{1.4} & \text{if } D_{\mathrm{b}} > 25.4\text{mm} \end{cases}.$$

Subject to $g_1(\boldsymbol{x}) = \frac{\phi_o}{2\sin^{-1}(D_b/D_m)} - Z + 1 \geq 0$, $g_2(\boldsymbol{x}) = 2D_b - K_{D\mathrm{min}}(D - d) \geq 0$, $g_3(\boldsymbol{x}) = K_{D\mathrm{max}}(D - d) - 2D_{\mathrm{b}} \geq 0$, $g_4(\boldsymbol{x}) = D_m - (0.5 - e)(D + d) \geq 0$, $g_5(\boldsymbol{x}) = (0.5 + e)(D + d) - D_m \geq 0$, $g_6(\boldsymbol{x}) = D_m - 0.5(D + d) \geq 0$, $g_7(\boldsymbol{x}) = 0.5(D - D_m - D_b) - \varepsilon D_b \geq 0$, $g_8(\boldsymbol{x}) = \zeta B_w - D_b \leq 0$, $g_9(\boldsymbol{x}) = f_i \geq 0.515$, $g_{10}(\boldsymbol{x}) = f_o \geq 0.515$.

where

$$f_c = 37.91 \left[1 + \left\{ 1.04 \left(\frac{1-\gamma}{1+\gamma} \right)^{1.72} \left(\frac{f_i(2f_o - 1)}{f_o(2f_i - 1)} \right)^{0.4} \right\}^{10/3} \right]^{-0.3}$$

$$\left(\frac{\gamma^{0.3}(1-\gamma)^{1.39}}{f_o(1+\gamma)^{\frac{1}{3}}} \right) \left(\frac{2f_i}{2f_i - 1} \right)^{0.41},$$

$\gamma = \frac{D_b \cos\alpha}{D_m}$, $f_i = \frac{r_i}{D_b}$, $f_o = \frac{r_o}{D_b}$,

$\phi_o = 2\pi - 2\cos^{-1}\frac{[(D-d)/2 - 3(T/4)]^2 + [D/2 - (T/4) - D_b]^2 - [d/2 + (T/4)]^2}{2[(D-d)/2 - 3(T/4)][D/2 - (T/4) - D_b]}$,

$T = D - d - 2D_b$, $D = 160$, $d = 90$, $B_w = 30$, $r_i = r_o = 11.033$.

Variable range $0.5(D + d) \leq D_m \leq 0.6(D + d)$, $0.15(D - d) \leq D_b \leq 0.45(D - d)$, $4 \leq Z \leq 50$, $0.515 \leq f_i \leq 0.6$, $0.515 \leq f_o \leq 0.6$, $0.4 \leq K_{D\mathrm{min}} \leq 0.5$, $0.6 \leq K_{D\mathrm{max}} \leq 0.7$, $0.3 \leq \varepsilon \leq 0.4$, $0.02 \leq e \leq 0.1$, $0.6 \leq \zeta \leq 0.85$.

B.9 Gear train design

Consider variable $\boldsymbol{x} = [x_1, x_2, x_3, x_4] = [T_a, T_b, T_d, T_f]$.

Minimize $f_9(\boldsymbol{x}) = \left(\frac{1}{6.931} + \frac{T_b \cdot T_d}{T_a \cdot T_f} \right)$.

Subject to (variable range) $0.01 \leq x_1, x_2, x_3, x_4 \leq 60$.

Appendix C Codes in MATLAB

The full codes with demonstrations and full description of each of three optimizations are downloadable from File Exchange at the MathWorks, Inc. website pages:

Atom Search Optimization (ASO): https://www.mathworks.com/matlabcentral/fileexchange/67011-atom-search-optimization-aso-algorithm.

Artificial Ecosystem-Based Optimization (AEO): https://www.mathworks.com/matlabcentral/fileexchange/72685-artificial-ecosystem-based-optimization-aeo.

Supply-Demand-Based Optimization (SDO): https://www.mathworks.com/matlabcentral/fileexchange/71764-supply-demand-based-optimization.

Index

Note: Page numbers followed by *f* indicate figures and *t* indicate tables.

D

E

F

G

H

I

L

M

N

O

P

T

W

Kolibri 360 GmbH
Ferdinand-Jühlke-Straße 7
99095 Erfurt, Deutschland
produktsicherheit@kolibri360.de